T0202704

Springer INdAM Series

Volume 25

More information about this series at http://www.springer.com/series/10283

Gabriele Bianchi • Andrea Colesanti •
Paolo Gronchi
Editors

Analytic Aspects
of Convexity

 Springer

Editors

Gabriele Bianchi
Dept. of Mathematics and Computer
Science "U. Dini"
University of Florence
Florence, Italy

Andrea Colesanti
Dept. of Mathematics and Computer
Science "U. Dini"
University of Florence
Florence, Italy

Paolo Gronchi
Dept. of Mathematics and Computer
Science "U. Dini"
University of Florence
Florence, Italy

ISSN 2281-518X   ISSN 2281-5198   (electronic)
Springer INdAM Series
ISBN 978-3-030-10121-3   ISBN 978-3-319-71834-7   (eBook)
https://doi.org/10.1007/978-3-319-71834-7

Printed on acid-free paper

This Springer imprint is published by Springer Nature
The registered company is Springer International Publishing AG
The registered company address is: Gewerbestrasse 11, 6330 Cham, Switzerland

Preface

This volume contains a selection of papers presented at the INdAM Workshop "Analytic Aspects of Convexity" and closely related works by some of the participants. This conference, held in Rome from October 10 to 14, 2016, represented a continuation of a series of conferences, with the same title, organized by Stefano Campi, Rolf Schneider and Aljoša Volčič. The previous conferences were held in Cortona every 4 years from 1995 to 2011.

Convexity is a notion which plays a significant role in numerous areas of mathematics and its use dates back several centuries. It was essentially the genius of Hermann Minkowski which, around the turn of the twentieth century, was responsible for the invention or discovery of that branch of mathematics that now goes under the name of Convex Geometry. Since then many mathematicians have explored and expanded the field using techniques from differential geometry, measure theory, combinatorics, algebraic geometry, Fourier analysis and probability. Interesting relations to other topics have also been disclosed, and, stealing a sentence from P.M. Gruber's History of Convexity (in: "Handbook of Convex Geometry", P.M. Gruber and J.M. Wills, editors, North-Holland Publishing Co., 1993), "part of the fascination of convexity is due to these interrelations". Nowadays, the research in Convex Geometry has many different souls and purposes. As an example we can refer to the books written by four participants in the conference: P.M. Gruber ("Convex and Discrete Geometry", Springer, 2007), R. Schneider ("Convex Bodies: the Brunn-Minkowski Theory", Cambridge University Press, 2014), A. Koldobsky ("Fourier Analysis in Convex Geometry", American Mathematical Society, 2005) and R.J. Gardner ("Geometric Tomography", Cambridge University Press, 2014).

The focus of the workshop was on some aspects of convexity where the connection with analysis, either in the methods or in the problems themselves, is particularly strong. The selection of papers in this volume offers a good reflection of the topics most represented in the meeting.

Geometric inequalities and problems of isoperimetric type are represented by the papers by M.A. Hernández Cifre and D. Alonso-Gutiérrez, A.E. Litvak, A. Stancu and J. Yepes Nicolás. The first of these papers deals with estimation of integrals of powers of the mean curvatures of the surface of a convex body in terms of geometric quantities of the body itself, like surface, volume or, more generally, intrinsic volumes. Litvak reports on an old conjecture regarding the simplex of maximum mean width among those inscribed in the unit ball, and the importance of this problem in probability theory and information theory. Stancu deals with problems in affine geometry and proposes a definition of centro-affine curvature and of affine length for polygons, with the latter satisfying isoperimetric type inequalities. The paper by J. Yepes Nicolás deals with the Brunn-Minkowski inequality, of central importance in convexity and in analysis.

Fourier analysis and geometric tomography are represented here by the paper by A. Koldobsky and D. Wu, which gives estimates on the difference in volume between two star bodies in terms of maximal or minimal difference between areas of sections of those bodies. These estimates are closely related to famous problems in geometric tomography, such as the slicing problem and the Busemann-Petty problem. Fourier analysis plays an essential role in the proofs of the estimates.

Integral geometry and the theory of valuations are represented by the papers by A. Bernig, J.H.G. Fu and G. Solanes and by D. Hug and J.A. Weis. The first paper deals with kinematic formulas and valuations on complex space forms, while Hug and Weis focus on integral geometry of tensor-valued generalisations of curvature measures of convex bodies.

Our thanks go to the Istituto Nazionale di Alta Matematica "Francesco Severi", which made it possible to hold the workshop through their financial support and by hosting us in their headquarters in Rome.

We had the honour of having Peter M. Gruber, who sadly passed away on March 7, 2017, as a participant and speaker in the meeting. He was a leading expert in the field and contributed deeply to the formation of this research community. We will all miss him greatly.

Florence, Italy   Gabriele Bianchi
October 2017   Andrea Colesanti
                                           Paolo Gronchi

Contents

About the Editors

Gabriele Bianchi, Andrea Colesanti and Paolo Gronchi are professors at the Department of Mathematics and Computer Science of the University of Florence. Their main research interest is in the analytic aspects of Convex Geometry.

Dual Curvature Measures in Hermitian Integral Geometry

Andreas Bernig, Joseph H. G. Fu, and Gil Solanes

Abstract The local kinematic formulas on complex space forms induce the structure of a commutative algebra on the space $\mathrm{Curv}^{\mathrm{U}(n)*}$ of dual unitarily invariant curvature measures. Building on the recent results from integral geometry in complex space forms, we describe this algebra structure explicitly as a polynomial algebra. This is a short way to encode all local kinematic formulas. We then characterize the invariant valuations on complex space forms leaving the space of invariant angular curvature measures fixed.

1 Introduction

Let $\mathbb{CP}^n_\lambda$ denote the complex space form of holomorphic sectional curvature 4λ and G_λ its holomorphic isometry group. By $\mathcal{C}(\mathbb{CP}^n_\lambda)^{G_\lambda}$ we denote the space of G_λ-invariant smooth curvature measures on $\mathbb{CP}^n_\lambda$ (see below for the definition).

The space $\mathcal{C}(\mathbb{CP}^n_\lambda)^{G_\lambda}$ is finite-dimensional, and several geometrically meaningful bases were used in [3]. Let $\Phi_1, \ldots, \Phi_m$ be such a basis. Then there are local kinematic formulas (cf. [4] or [7])

$$\int_{G_\lambda} \Phi_j(P_1 \cap gP_2, \beta_1 \cap g\beta_2)dg = \sum_{k,l} c^j_{k,l}\Phi_k(P_1, \beta_1)\Phi_l(P_2, \beta_2).$$

A. Bernig (✉)
Institut für Mathematik, Goethe-Universität Frankfurt, Frankfurt, Germany
e-mail: bernig@math.uni-frankfurt.de

J. H. G. Fu
Department of Mathematics, University of Georgia, Athens, GA, USA
e-mail: fu@math.uga.edu

G. Solanes
Departament de Matemàtiques, Universitat Autònoma de Barcelona, Bellaterra, Spain
e-mail: solanes@mat.uab.cat

© Springer International Publishing AG 2018
G. Bianchi et al. (eds.), *Analytic Aspects of Convexity*, Springer INdAM Series 25,
https://doi.org/10.1007/978-3-319-71834-7_1

1

Here P_1, P_2 are compact submanifolds with corners and β_1, β_2 are Borel subsets of $\mathbb{CP}^n_\lambda$. Finding the constants $c^j_{k,l}$ is a non-trivial question which could be solved only recently [2, 3].

By the transfer principle [3], the spaces $\mathcal{C}(\mathbb{CP}^n_\lambda)^{G_\lambda}$ are naturally identified with the space $\mathrm{Curv}^{\mathrm{U}(n)}$ of smooth, translation and $\mathrm{U}(n)$-invariant curvature measures on the hermitian space $\mathbb{C}^n$ and under this identification, the local kinematic formulas are independent of λ. We may therefore define an operator

$$K : \mathrm{Curv}^{\mathrm{U}(n)} \to \mathrm{Curv}^{\mathrm{U}(n)} \otimes \mathrm{Curv}^{\mathrm{U}(n)}, \quad \Phi_j \mapsto \sum_{k,l} c^j_{k,l} \Phi_k \otimes \Phi_l,$$

which makes $\mathrm{Curv}^{\mathrm{U}(n)}$ into a cocommutative, coassociative coalgebra.

Stated otherwise, the dual space $\mathrm{Curv}^{\mathrm{U}(n)*}$ becomes a commutative associative algebra with respect to the product

$$K^* : \mathrm{Curv}^{\mathrm{U}(n)*} \otimes \mathrm{Curv}^{\mathrm{U}(n)*} \to \mathrm{Curv}^{\mathrm{U}(n)*}.$$

The knowledge of the algebra structure on $\mathrm{Curv}^{\mathrm{U}(n)*}$ is equivalent to the knowledge of the local kinematic formulas. Hence the description of this structure is a short and elegant way of stating the local kinematic formulas. This will be achieved by the first main theorem of this paper.

Before stating it, let us introduce some notation and recall a result by Fu [5]. Let $\mathrm{Val}^{\mathrm{U}(n)}$ be the space of continuous, translation invariant and $\mathrm{U}(n)$-invariant valuations (see Sect. 2 for the definition of valuations). Let $\phi_1, \ldots, \phi_r$ be a basis of $\mathrm{Val}^{\mathrm{U}(n)}$. Then there are global kinematic formulas

$$\int_{\overline{\mathrm{U}(n)}} \phi_j(P_1 \cap \bar{g} P_2) d\bar{g} = \sum_{k,l} \tilde{c}^j_{k,l} \phi_k(P_1) \phi_l(P_2).$$

The operator $k : \mathrm{Val}^{\mathrm{U}(n)} \to \mathrm{Val}^{\mathrm{U}(n)} \otimes \mathrm{Val}^{\mathrm{U}(n)}$ defined by $\phi_j \mapsto \sum_{k,l} \tilde{c}^j_{k,l} \phi_k \otimes \phi_l$ induces the structure of a commutative, associative algebra

$$k^* : \mathrm{Val}^{\mathrm{U}(n)^*} \otimes \mathrm{Val}^{\mathrm{U}(n)^*} \to \mathrm{Val}^{\mathrm{U}(n)^*}.$$

In this case, we may identify $\mathrm{Val}^{\mathrm{U}(n)}$ and $\mathrm{Val}^{\mathrm{U}(n)^*}$ by the Alesker-Poincaré pairing and the resulting algebra structure on $\mathrm{Val}^{\mathrm{U}(n)}$ is the Alesker product of valuations defined in [1].

Theorem 1.1 ([5]) *The algebra* $\mathrm{Val}^{\mathrm{U}(n)}$ *is given by*

$$\mathrm{Val}^{\mathrm{U}(n)} \cong \mathbb{C}[t, s]/(f_{n+1}, f_{n+2}),$$

where t, s are variables of degrees 1 and 2 respectively, and f_k is the part of total degree k in the series expansion of $\log(1 + t + s)$.

Let t, s, v be variables of degrees 1, 2 and 3 respectively and $u := 4s - t^2$. Wannerer [9] defined the polynomial q_n as the n-homogeneous part in the expansion of $-\frac{1}{(1+t+s)^2}$. These polynomials appear in the description of the algebra $\text{Area}^{U(n)*}$ of unitarily invariant dual area measures [8, 9].

Our first main theorem gives an algebraic way of encoding all local kinematic formulas, and, moreover, answers [3, Question 7.3].

Theorem 1.2 *There is an algebra isomorphism*

$$\text{Curv}^{U(n)*} \cong \mathbb{C}[t, s, v]/(f_{n+1}, f_{n+2}, q_{n-1}v, q_n v, (v + tu)^2).$$

Let us now describe our second main theorem. Let M be a Riemannian manifold. The space $\mathcal{V}(M)$ of smooth valuations on M admits an algebra structure, and the space $\mathcal{C}(M)$ of smooth curvature measures on M is a module over this algebra [3]. In [3] we defined the subspace $\mathcal{A}(M) \subset \mathcal{C}(M)$ of *angular curvature measures*. A smooth valuation μ on M with the property that $\mu \cdot \mathcal{A}(M) \subset \mathcal{A}(M)$ is called *angular valuation*; the corresponding space is denoted by $\mathfrak{a}(M) \subset \mathcal{V}(M)$. The Lipschitz-Killing algebra of M [3, Subsection 2.7] is denoted by $\text{LK}(M)$. Conjecture 3 of [3] states that for every Riemannian manifold M we have

$$\mathfrak{a}(M) = \text{LK}(M).$$

Neither of the inclusions seems to be known.

We study a version of this conjecture on $M = \mathbb{CP}^n_\lambda, \lambda \in \mathbb{R}$. More precisely, we characterize the space of all invariant valuations on $\mathbb{CP}^n_\lambda$ which leave the space of *invariant* angular curvature measures invariant. As in the flat case, the algebra $\mathcal{V}^n_\lambda$ of G_λ-invariant valuations in $\mathbb{CP}^n_\lambda$ has two natural generators: t_λ and s (cf. [3]).

Theorem 1.3 *The algebra of G_λ-invariant valuations on $\mathbb{CP}^n_\lambda$ leaving the space of angular invariant curvature measures invariant is given by the elements $p(t_\lambda, s) \in \mathcal{V}^n_\lambda$, with $p \in \mathbb{C}[t, s]$ such that*

$$tsu\frac{\partial p}{\partial s}\left(\frac{t}{(1 - \lambda s)^{\frac{1}{2}}}, s\right) = 0 \text{ in } \text{Val}^{U(n)}.$$

If p is a polynomial in t_λ alone, then clearly p satisfies the angularity condition. However, since $\text{Val}^{U(n)}$ contains zero divisors, the equation in the corollary does not imply that $\frac{\partial p}{\partial s} = 0$ in general. For instance, the image of the polynomial $p(t, s) = t^4 - 6st^2 + 6s^2$ is angular in $\text{Val}^{U(4)}$, since $tsu(-6t^2 + 12s) = 0$. However, p can not be written as a polynomial in t alone, since there are no relations between t and s of degree 4.

2 Background and Notations

For the reader's convenience, we collect here some results from [3] which will be needed in the sequel.

Let M be a smooth manifold of dimension n which for simplicity is assumed to be oriented and connected. The space of compact submanifolds with corners is denoted by $\mathcal{P}(M)$. A smooth curvature measure is a functional of the form

$$\Phi(P, \beta) := \int_{N(P) \cap \pi^{-1}\beta} \omega + \int_{P \cap \beta} \rho,$$

where $N(P)$ is the conormal cycle of P (which is a Legendrian cycle in the cosphere bundle $S^*(M)$); β a Borel subset of M; $\omega \in \Omega^{n-1}(S^*M)$, $\pi : S^*(M) \to M$ the natural projection and $\rho \in \Omega^n(M)$. The space of smooth curvature measures is denoted by $\mathcal{C}(M)$.

A functional of the form

$$\phi(P) := \int_{N(P)} \omega + \int_{P} \rho,$$

is called smooth valuation on M. The corresponding space is denoted by $\mathcal{V}(M)$. The obvious map glob : $\mathcal{C}(M) \to \mathcal{V}(M)$ is called globalization map.

Alesker [1] has introduced a product structure on $\mathcal{V}(M)$, which was generalized in [3] to a module structure of $\mathcal{C}(M)$ over $\mathcal{V}(M)$ satisfying

$$\text{glob}(\phi \cdot \Phi) = \phi \cdot \text{glob } \Phi, \quad \phi \in \mathcal{V}(M), \Phi \in \mathcal{C}(M).$$

In the special case where $M = V$ is an affine space, we let Curv = Curv(V) be the space of translation invariant smooth curvature measures and $\text{Val}^\infty = \text{Val}^\infty(V)$ the space of translation invariant smooth valuations. Then Curv is a module over Val^∞.

Let us now specialize to the case $\text{Curv}^{U(n)}$ of unitarily invariant elements in Curv. As a vector space, it is generated by elements $\Delta_{k,q}, 0 \le k \le 2n, \max\{0, k - n\} \le q \le \lfloor \frac{k}{2} \rfloor$ and $N_{k,q}, 1 \le k \le 2n - 3, \max\{0, k - n + 1\} \le q < \frac{k}{2}$. They are defined in terms of invariant differential forms on the sphere bundle, we refer to [3, Subsection 3.1] for details. The globalization map is injective on $\text{span}\{\Delta_{k,q}\}$ while its kernel is spanned by $\{N_{k,q}\}$.

The algebra $\text{Val}^{U(n)}$ is generated by two elements t, s. The module structure of $\text{Curv}^{U(n)}$ over $\text{Val}^{U(n)}$ was computed in [3] as follows.

$$s \cdot \Delta_{kq} = \frac{(k - 2q + 2)(k - 2q + 1)}{2\pi(k + 2)} \Delta_{k+2,q} + \frac{2(q + 1)(k - q + 1)}{\pi(k + 2)} \Delta_{k+2,q+1}$$

$$- \frac{(k - 2q + 2)(k - 2q + 1)}{\pi(k + 2)(k + 4)} N_{k+2,q} - \frac{2(q + 1)(k - 2q)}{\pi(k + 2)(k + 4)} N_{k+2,q+1}, \quad (1)$$

$$s \cdot N_{kq} = \frac{(k - 2q + 2)(k - 2q + 1)}{2\pi(k + 4)} N_{k+2,q} + \frac{2(q + 1)(k - q + 2)}{\pi(k + 4)} N_{k+2,q+1},$$

(2)

$$t \cdot \Delta_{kq} = \frac{\omega_{k+1}}{\pi\omega_k} \left((k - 2q + 1)\Delta_{k+1,q} + 2(q + 1)\Delta_{k+1,q+1} \right),$$

(3)

$$t \cdot N_{kq} = \frac{\omega_{k+1}}{\pi\omega_k} \frac{k + 2}{k + 3} \left((k - 2q + 1)N_{k+1,q} + \frac{2(q + 1)}{k - 2q}(k - 2q - 1)N_{k+1,q+1} \right)$$

(4)

where ω_i denotes the volume of the i-dimensional unit ball.

The map $\phi \mapsto \phi \cdot \Delta_{0,0}$ is called the ƒ-map, while the map $\phi \mapsto \phi \cdot N_{1,0}$ is called the n-map. Then $\mathrm{Curv}^{U(n)}$ is generated by the images of the ƒ- and n-maps. More precisely,

$$\mathrm{Curv}^{U(n)} = \mathrm{Val}^{U(n)} \Delta_{0,0} \oplus \widetilde{\mathrm{Val}}^{U(n)} N_{1,0},$$

(5)

where $\widetilde{\mathrm{Val}}^{U(n)} := \mathbb{C}[t, s]/(q_{n-1}, q_n)$ is a graded quotient of $\mathrm{Val}^{U(n)}$. More precisely, it was shown in [3] that $\widetilde{\mathrm{Val}}^{U(n)} := \mathbb{C}[t, s]/(g_{n-1}, g_n)$, where g_n is the degree n-part in

$$e^t \frac{\sin\sqrt{u} - \sqrt{u}\cos\sqrt{u}}{2\sqrt{u}^3}.$$

By [9, Lemma 4.4] and [3, Proposition 5.15], we have

$$q_n = \frac{(-1)^{n+1}(n + 3)!}{2^n} g_n,$$

so that the two descriptions of $\widetilde{\mathrm{Val}}^{U(n)}$ agree.

The algebra of G_λ-invariant smooth valuations on $\mathbb{CP}_\lambda^n$ is denoted by $\mathcal{V}_\lambda^n$. It is generated by the generator t_λ of the Lipschitz-Killing algebra of $\mathbb{CP}_\lambda^n$ and another element denoted by s [3].

By the transfer principle, the space $\mathcal{C}(\mathbb{CP}_\lambda^n)^{G_\lambda}$ of invariant smooth curvature measures on $\mathbb{CP}_\lambda^n$ can be naturally identified with $\mathrm{Curv}^{U(n)}$. The local kinematic formulas on the different $\mathbb{CP}_\lambda^n$ are then formally identical. There is a module structure on $\mathcal{C}(\mathbb{CP}_\lambda^n)^{G_\lambda}$ over $\mathcal{V}_\lambda^n$ which is closely related to the so-called semi-local formulas. These are maps

$$\bar{k}_\lambda : \mathrm{Curv}^{U(n)} \to \mathrm{Curv}^{U(n)} \otimes \mathcal{V}_\lambda^n$$

$$\Phi \mapsto \left[(P_1, \beta, P_2) \mapsto \int_{G_\lambda} \Phi(P_1 \cap gP_2, \beta)dg \right].$$

Writing $\mathrm{glob}_\lambda : \mathrm{Curv}^{\mathrm{U}(n)} \to \mathcal{V}^n_\lambda$ for the globalization map on $\mathbb{CP}^n_\lambda$, we obviously have $\bar{k}_\lambda = (\mathrm{id} \otimes \mathrm{glob}_\lambda) \circ K$. The global kinematic formulas on $\mathcal{V}^n_\lambda$ are given by $k_\lambda \circ \mathrm{glob}_\lambda = (\mathrm{glob}_\lambda \otimes \mathrm{glob}_\lambda) \circ K$; i.e.

$$k_\lambda : \mathcal{V}^n_\lambda \to \mathcal{V}^n_\lambda \otimes \mathcal{V}^n_\lambda$$

$$\phi \mapsto \left[(P_1, P_2) \mapsto \int_{G_\lambda} \phi(P_1 \cap gP_2)dg \right]$$

Let $\mathrm{pd}_\lambda : \mathcal{V}^n_\lambda \to \mathcal{V}^{n*}_\lambda$ be the normalized Poincaré duality, see [3, Definition 2.14]. We will use the notation $\mathrm{PD} := \mathrm{pd}_0 : \mathrm{Val}^{\mathrm{U}(n)} \to \mathrm{Val}^{\mathrm{U}(n)*}$. This map satisfies $\langle \mathrm{PD}\phi, \psi \rangle = (\phi \cdot \psi)_{2n}$, where $\phi \cdot \psi$ is the Alesker product and the subindex $2n$ denotes the component of highest degree.

The module product of $\mathcal{V}^n_\lambda$ on $\mathrm{Curv}^{\mathrm{U}(n)}$ is a map $\bar{m}_\lambda \in \mathrm{Hom}(\mathcal{V}^n_\lambda \otimes \mathrm{Curv}^{\mathrm{U}(n)}, \mathrm{Curv}^{\mathrm{U}(n)})$, which turns out to be equivalent to the semi-local kinematic formulas by

$$\bar{m}_\lambda = (\mathrm{id} \otimes \mathrm{pd}_\lambda) \circ \bar{k}_\lambda, \tag{6}$$

under the identification

$$\mathrm{Hom}(\mathcal{V}^n_\lambda \otimes \mathrm{Curv}^{\mathrm{U}(n)}, \mathrm{Curv}^{\mathrm{U}(n)}) = \mathrm{Hom}(\mathrm{Curv}^{\mathrm{U}(n)}, \mathrm{Curv}^{\mathrm{U}(n)} \otimes \mathcal{V}^{n*}_\lambda).$$

Similarly, the Alesker product is a map $m_\lambda : \mathcal{V}^n_\lambda \otimes \mathcal{V}^n_\lambda \to \mathcal{V}^n_\lambda$ which satisfies $(\mathrm{pd}_\lambda \otimes \mathrm{pd}_\lambda) \circ k_\lambda = m^*_\lambda \circ \mathrm{pd}_\lambda$.

We may summarize the different maps in the following commuting diagram.

$$
\begin{array}{ccc}
\mathrm{Curv}^{\mathrm{U}(n)} & \xrightarrow{\;\;K\;\;} & \mathrm{Curv}^{\mathrm{U}(n)} \otimes \mathrm{Curv}^{\mathrm{U}(n)} \\
\downarrow{\scriptstyle \mathrm{id}} & & \downarrow{\scriptstyle \mathrm{id}\otimes\mathrm{glob}_\lambda} \\
\mathrm{Curv}^{\mathrm{U}(n)} & \xrightarrow{\;\;\bar{k}_\lambda\;\;} & \mathrm{Curv}^{\mathrm{U}(n)} \otimes \mathcal{V}^n_\lambda \\
\downarrow{\scriptstyle \mathrm{glob}_\lambda} & & \downarrow{\scriptstyle \mathrm{glob}_\lambda \otimes \mathrm{id}} \\
\mathcal{V}^n_\lambda & \xrightarrow{\;\;k_\lambda\;\;} & \mathcal{V}^n_\lambda \otimes \mathcal{V}^n_\lambda \\
\downarrow{\scriptstyle \mathrm{pd}_\lambda} & & \downarrow{\scriptstyle \mathrm{pd}_\lambda \otimes \mathrm{pd}_\lambda} \\
\mathcal{V}^{n*}_\lambda & \xrightarrow{\;\;m^*_\lambda\;\;} & \mathcal{V}^{n*}_\lambda \otimes \mathcal{V}^{n*}_\lambda
\end{array}
\tag{7}
$$

3 The Algebra Structure on $\mathrm{Curv}^{U(n)*}$

In this section, we will prove Theorem 1.2.

Since local and global kinematic formulas are intertwined by glob_λ, the maps $\mathrm{glob}_\lambda^* : \mathcal{V}_\lambda^{n*} \to \mathrm{Curv}^{U(n)*}$ and $\mathrm{glob}_\lambda^* \circ \mathrm{pd}_\lambda : \mathcal{V}_\lambda^n \to \mathrm{Curv}^{U(n)*}$ are injective algebra morphisms. Given an element $\phi \in \mathcal{V}_\lambda^n$, we will denote by $\bar{\phi}$ its image in $\mathrm{Curv}^{U(n)*}$. In particular, the elements $t, s \in \mathrm{Val}^{U(n)}$ give rise to elements $\bar{t}, \bar{s} \in \mathrm{Curv}^{U(n)*}$.

Let us denote by $\Delta_{k,q}^*, N_{k,q}^* \in \mathrm{Curv}^{U(n)*}$ the dual basis of the basis $\Delta_{k,q}, N_{k,q}$ of $\mathrm{Curv}^{U(n)}$. The unit element in Val is the Euler characteristic χ. The unit element in $\mathrm{Curv}^{U(n)*}$ is $\bar{\chi} = \Delta_{2n,n}^*$.

Proposition 3.1 *Let $\phi \in \mathcal{V}_\lambda^n$ and $\bar{m}_\phi : \mathrm{Curv}^{U(n)} \to \mathrm{Curv}^{U(n)}$ be the module multiplication by ϕ. Then the dual map $\bar{m}_\phi^* : \mathrm{Curv}^{U(n)*} \to \mathrm{Curv}^{U(n)*}$ equals multiplication by $\bar{\phi}$.*

Proof Let $L \in \mathrm{Curv}^{U(n)*}$ and $\Phi \in \mathrm{Curv}^{U(n)}$. By (6),

$$
\begin{aligned}
\langle \bar{m}_\phi^* L, \Phi \rangle &= \langle L, \bar{m}_\phi(\Phi) \rangle \\
&= \langle L \otimes \mathrm{pd}_\lambda(\phi), \bar{k}_\lambda(\Phi) \rangle \\
&= \langle L \otimes \mathrm{pd}_\lambda(\phi), (\mathrm{id} \otimes \mathrm{glob}_\lambda) \circ K(\Phi) \rangle \\
&= \langle L \otimes (\mathrm{glob}_\lambda^* \circ \mathrm{pd}_\lambda)(\phi), K(\Phi) \rangle \\
&= \langle K^*(L \otimes \bar{\phi}), \Phi \rangle \\
&= \langle L \cdot \bar{\phi}, \Phi \rangle.
\end{aligned}
$$

$\square$

Lemma 3.2

$$
\bar{t}\Delta_{k,q}^* = \frac{\omega_k}{\pi \omega_{k-1}} \left[(k - 2q)\Delta_{k-1,q}^* + 2q\Delta_{k-1,q-1}^* \right],
$$

$$
\bar{t}N_{k,q}^* = \frac{\omega_k}{\pi \omega_{k-1}} \frac{k+1}{k+2} \left[(k - 2q)N_{k-1,q}^* + \frac{2q(k-2q)}{k-2q+1}N_{k-1,q-1}^* \right],
$$

$$
\bar{s}\Delta_{k,q}^* = \frac{(k-2q)(k-2q-1)}{2\pi k}\Delta_{k-2,q}^* + \frac{2q(k-q)}{\pi k}\Delta_{k-2,q-1}^*,
$$

$$
\begin{aligned}
\bar{s}N_{k,q}^* = &-\frac{(k-2q)(k-2q-1)}{\pi k(k+2)}\Delta_{k-2,q}^* - \frac{2q(k-2q)}{\pi k(k+2)}\Delta_{k-2,q-1}^* \\
&+ \frac{(k-2q)(k-2q-1)}{2\pi(k+2)}N_{k-2,q}^* + \frac{2q(k-q+1)}{\pi(k+2)}N_{k-2,q-1}^*.
\end{aligned}
$$

Proof These equations follow from (1) to (4) and Proposition 3.1. $\square$

In particular, $\bar{t} = \frac{2n\omega_{2n}}{\pi\omega_{2n-1}}\Delta^*_{2n-1,n-1}$ and $\bar{s} = \frac{n}{\pi}\Delta^*_{2n-2,n-1}$. It will be also useful to compute

$$\bar{t}\bar{s} = \frac{4\omega_{2n}(n-1)n}{\omega_{2n-1}(2n-1)\pi^2}n\Delta^*_{2n-3,n-2} \tag{8}$$

$$\bar{t}^3 = \frac{4\omega_{2n}(n-1)n}{\omega_{2n-1}(2n-1)\pi^2}\left(4(n-2)\Delta^*_{2n-3,n-3} + 6\Delta^*_{2n-3,n-2}\right). \tag{9}$$

Besides $\bar{t}, \bar{s}$ we will need a third element in $\mathrm{Curv}^{U(n)*}$, which will be denoted by $\bar{v}$ (even though it is not the image of an element $v \in \mathrm{Val}^{U(n)}$). Namely,

$$\bar{v} := \frac{16\omega_{2n}n(n-1)(n-2)}{\omega_{2n-1}(2n-1)\pi^2}\left(\Delta^*_{2n-3,n-3} - \Delta^*_{2n-3,n-2} - \frac{2n-1}{2(n-2)}N^*_{2n-3,n-2}\right). \tag{10}$$

Proof of Theorem 1.2 The image of the $\mathfrak{l}$-map in $\mathrm{Curv}^{U(n)}_{2n-3}$ has dimension 2, while $\dim\mathrm{Curv}^{U(n)}_{2n-3} = 3$. We claim that $\bar{v}$ vanishes on the image of the $\mathfrak{l}$-map, which defines $\bar{v}$ uniquely (up to scale).

By [3, Lemma 5.12], and (1),

$$\mathfrak{l}(t^{2n-3}) = \frac{2^{2n-3}(n-2)!}{\pi^{n-1}}(\Delta_{2n-3,n-3} + \Delta_{2n-3,n-2}) \tag{11}$$

$$\mathfrak{l}(t^{2n-5}u) = \frac{2^{2n-3}(n-2)!}{\pi^{n-1}(2n-3)}\left(\frac{n-3}{n-2}\Delta_{2n-3,n-3} + \Delta_{2n-3,n-2} - \frac{2}{2n-1}N_{2n-3,n-2}\right), \tag{12}$$

from which the claim follows.

Using reverse induction on the degree k and Lemma 3.2, one can show that each element $L \in \mathrm{Curv}^*$ may be written as

$$L = p(\bar{s}, \bar{t}) + q(\bar{s}, \bar{t})\bar{v} \tag{13}$$

with polynomials $p, q \in \mathbb{C}[\bar{s}, \bar{t}]$. In particular, $\bar{s}, \bar{t}, \bar{v}$ generate the algebra $\mathrm{Curv}^{U(n)*}$.

It follows from [5] and [3, Proposition 5.15] that if p is in the ideal generated by $f_{n+1}(\bar{s}, \bar{t}), f_{n+2}(\bar{s}, \bar{t})$ and q is in the ideal generated by $q_{n-1}(\bar{s}, \bar{t})$ and $q_n(\bar{s}, \bar{t})$ then $L = p(\bar{s}, \bar{t}) + q(\bar{s}, \bar{t})\bar{v} = 0$. Indeed, for $\Phi = \mu\Delta_{0,0} + \varphi N_{1,0} \in \mathrm{Curv}^{U(n)}$,

$$\langle L, \Phi \rangle = \langle 1, p(s,t)\Phi \rangle + \langle \bar{v}, q(s,t)\mu\Delta_{0,0} \rangle + \langle \bar{v}, q(s,t)\varphi N_{1,0} \rangle = 0.$$

By looking at the dimensions, one sees that there can be no more relations of degree 1 in $\bar{v}$. This fixes the algebra structure on $\mathrm{Curv}^{U(n)*}$, except that we have to write $\bar{v}^2$ in the form (13).

There is an algebra isomorphism $I_\lambda : \mathrm{Val}^{U(n)} \to \mathcal{V}^n_\lambda$ [3, Thm. 3.17] defined by $t \mapsto t_\lambda \sqrt{1 - \lambda s}, s \mapsto s$. The map $H_\lambda := \mathrm{PD}^{-1} \circ I^*_\lambda \circ \mathrm{pd}_\lambda \circ \mathrm{glob}_\lambda : \mathrm{Curv}^{U(n)} \to \mathrm{Val}^{U(n)}$ was studied in [3, Section 6]. With $H'_0 := \frac{d}{d\lambda}|_{\lambda=0} H_\lambda$ we have $H'_0 \circ \mathfrak{l} = D_1, H'_0 \circ \mathfrak{n} = D_2$, where $D_1, D_2 : \mathrm{Val}^{U(n)} \to \mathrm{Val}^{U(n)}$ are covered by the maps

$$D_1 p := \frac{t^2 - 2s}{2} p - \frac{tu}{4} \frac{\partial p}{\partial t}, \quad D_2 p := -\frac{3\pi ut}{8} p + \frac{\pi u^2}{8} \frac{\partial p}{\partial t}, p \in \mathbb{C}[t, s].$$

Using

$$t^{2n-3} s = \frac{n}{2(2n-1)} t^{2n-1}, \quad t^{2n-5} s^2 = \frac{n(n-1)}{4(2n-1)(2n-3)} t^{2n-1}$$

in $\mathrm{Val}^{U(n)}$, one easily computes that

$$\mathfrak{l}(t^{2n-3}) + \frac{2(2n-3)}{3\pi} \mathfrak{n}(t^{2n-4}) \in \ker H'_0, \tag{14}$$

$$\mathfrak{l}(t^{2n-5}u) + \mathfrak{n}\left(\frac{2}{\pi} t^{2n-4}\right) \in \ker H'_0. \tag{15}$$

By [3, Lemma 5.12.], we have

$$\mathfrak{n}(t^{2n-4}) = \frac{3}{4} \frac{(2n-4)!\omega_{2n-1}}{\pi^{2n-3}} N_{2n-3,n-2}. \tag{16}$$

Using (11), (12) and (16), we see that (14) and (15) are equivalent to

$$\Delta_{2n-3,n-3} + \Delta_{2n-3,n-2} + \frac{1}{2n-1} N_{2n-3,n-2} \in \ker H'_0 \tag{17}$$

$$\frac{n-3}{n-2} \Delta_{2n-3,n-3} + \Delta_{2n-3,n-2} + \frac{1}{2n-1} N_{2n-3,n-2} \in \ker H'_0. \tag{18}$$

It follows that

$$\Delta^*_{2n-3,n-2} - (2n-1)N^*_{2n-3,n-2} \in \mathrm{Im}(H'_0)^*.$$

This element is (up to a scaling factor) just $\bar{v} + \bar{t}u$ (see (8), (9)). By [3, Prop. 6.2.], we have $(H'_0 \otimes H'_0) \circ K = 0$. Dualizing, we obtain that the restriction of the product to $\mathrm{Im}(H'_0)^*$ vanishes. In particular, the square of $\bar{v} + \bar{t}u$ vanishes. $\qquad \square$

We may now complete the description of the product structure on $\mathrm{Curv}^{U(n)*}$ given in Lemma 3.2.

Lemma 3.3

$$\bar{v}\Delta^*_{k,q} = \frac{16}{\pi^2}\frac{\omega_k}{\omega_{k-1}(k-1)}\left[6\binom{q}{3}\Delta^*_{k-3,q-3} + \binom{q}{2}(k-4q+4)\Delta^*_{k-3,q-2}\right.$$

$$\left. -\binom{q}{2}(k-2q)\Delta^*_{k-3,q-1} - \frac{k-1}{k-2q+1}\binom{q}{2}N^*_{k-3,q-2}\right],$$

$$\bar{v}N^*_{k,q} = \frac{\omega_k(k-2q)}{\pi^2\omega_{k-1}(k+2)}\left[\frac{(k-2q-1)(k-2q-2)}{k-1}\Delta^*_{k-3,q}\right.$$

$$+ 12\frac{(2k-4q-1)q}{k-1}\Delta^*_{k-3,q-1} + 24\frac{q(q-1)}{k-1}\Delta^*_{k-3,q-2}$$

$$+ 32\frac{q-2}{k-2q+3}\binom{q}{2}N^*_{k-3,q-3}$$

$$+ 16\frac{k-4q-3}{k-2q+1}\binom{q}{2}N^*_{k-3,q-2}$$

$$\left. - 16\frac{q+2}{q-1}\binom{q}{2}N^*_{k-3,q-1}\right].$$

Proof There can be only one linear operator σ of degree -3 acting on $\mathrm{Curv}^{U(n)*}$ with the following properties: σ commutes with multiplications by $\bar{t}$ and $\bar{s}$, $\sigma\Delta^*_{2n,n} = \bar{v}$ and

$$\sigma\bar{v} + 2\bar{t}\bar{u}\bar{v} + (\bar{t}\bar{u})^2 = 0. \tag{19}$$

Indeed, for polynomials p_1, p_2 in $\bar{t}, \bar{s}$, $\sigma(p_1 + p_2\bar{v}) = p_1\sigma\Delta^*_{2n,n} + p_2\sigma\bar{v}$ is uniquely determined by these properties. The multiplication by $\bar{v}$ has these properties, where (19) follows from $(\bar{v} + \bar{t}\bar{u})^2 = 0$ which is true by Theorem 1.2. The operator on the right hand side of the displayed equations also has these properties (which is a bit tedious to verify), and hence both sides of the displayed equation agree. $\square$

4 The Image of $\mathcal{V}^n_\lambda$ in $\mathrm{Curv}^{U(n)*}$

Recall the injection of algebras $\mathcal{V}^n_\lambda \longrightarrow \mathrm{Curv}^{U(n)*}$ given by $\phi \mapsto \bar{\phi} = \mathrm{glob}^*_\lambda \circ \mathrm{pd}_\lambda(\phi)$. Since $\mathcal{V}^n_\lambda$ is generated by s and t_λ, we can describe this morphism by finding the images of s and t_λ. Since module multiplication by s is independent of the curvature (see [3, Prop.5.2]), it follows that $s \in \mathcal{V}^n_\lambda$ is mapped to the same element $\bar{s} \in \mathrm{Curv}^{U(n)*}$ for all λ. It remains to find the image of t_λ.

Lemma 4.1 *For all $p, q \in \mathbb{C}[t, s]$ we have*

$$\langle \bar{q}, \mathfrak{l}(p) \rangle = \langle \mathrm{PD}(q), p \rangle \tag{20}$$

and

$$\langle \bar{v}, \mathfrak{n}(p) \rangle = \langle \mathrm{PD}(p), e \rangle \tag{21}$$

where $e := -\frac{\pi}{2} u^2$.

Proof The first equation follows from

$$
\begin{aligned}
\langle \bar{q}, \mathfrak{l}(p) \rangle &= \langle \mathrm{glob}^* \circ \mathrm{PD}(q), \mathfrak{l}(p) \rangle \\
&= \langle \mathrm{PD}(q), \mathrm{glob} \circ \mathfrak{l}(p) \rangle \\
&= \langle \mathrm{PD}(q), p \rangle.
\end{aligned}
$$

For the second equation, we note that both sides vanish if p is not of degree $2n - 4$. We may thus suppose that p is a linear combination of $t^{2n-4}, t^{2n-6}u, t^{2n-8}u^2$. Using [3, Lemma 5.12] and [2, Prop. 3.7] this is a tedious, but straightforward computation. $\qquad \square$

Proposition 4.2 *The image of t_λ in $\mathrm{Curv}^{U(n)*}$ is given by*

$$\bar{t}_\lambda = \frac{\bar{t} - \frac{\lambda \bar{t}^3}{4}}{(1 - \lambda \bar{s})^{\frac{3}{2}}} + \frac{\lambda}{4(1 - \lambda \bar{s})^{\frac{3}{2}}} \bar{v}.$$

Proof Let $p_1 := \frac{\bar{t} - \frac{\lambda \bar{t}^3}{4}}{(1 - \lambda \bar{s})^{\frac{3}{2}}}, p_2 := \frac{\lambda}{4(1 - \lambda \bar{s})^{\frac{3}{2}}}$. We have to show that $\bar{t}_\lambda = p_1 + p_2 \bar{v}$.
By [3, Thm. 6.7] we have

$$t_\lambda \Delta_{0,0} = p_1 \Delta_{0,0} + r_1 N_{1,0},$$

$$t_\lambda N_{1,0} = p_2 e \Delta_{0,0} + r_2 N_{1,0},$$

where r_1, r_2 are explicitly known but irrelevant for our purpose.
Let $q_1 \in \mathrm{Val}^{U(n)}$. Then by (20)

$$\langle p_1 + p_2 \bar{v}, q_1 \Delta_{0,0} \rangle = \langle p_1, q_1 \Delta_{0,0} \rangle = \langle \mathrm{PD}(p_1), q_1 \rangle$$

and

$$\langle \bar{t}_\lambda, q_1 \Delta_{0,0} \rangle = \langle \Delta_{2n,n}^*, t_\lambda q_1 \Delta_{0,0} \rangle = \langle \Delta_{2n,n}^*, q_1 (p_1 \Delta_{0,0} + r_1 N_{1,0}) \rangle = \langle \mathrm{PD}(p_1), q_1 \rangle.$$

Next we use (21) and compute, for $q_2 \in \mathrm{Val}^{U(n)}$

$$\langle p_1 + p_2 \bar{v}, q_2 N_{1,0} \rangle = \langle p_2 \bar{v}, q_2 N_{1,0} \rangle = \langle \bar{v}, p_2 q_2 N_{1,0} \rangle = \langle \mathrm{PD}(p_2 q_2), e \rangle$$

and

$$\langle t_\lambda, q_2 N_{1,0} \rangle = \langle q_2 t_\lambda N_{1,0}, \Delta^*_{2n,n} \rangle = \langle q_2 (p_2 e \Delta_{0,0} + r_2 N_{1,0}), \Delta^*_{2n,n} \rangle = \langle \mathrm{PD}(p_2 q_2), e \rangle.$$

It follows that $\bar{t}_\lambda$ and $p_1 + p_2 \bar{v}$ act the same on the images of the l- and n-maps, hence on all $\mathrm{Curv}^{U(n)}$, which finishes the proof.   □

Next we want to describe the image of $\mathcal{V}_n^\lambda$ in $\mathrm{Curv}^{U(n)*}$.

Lemma 4.3 *Let $p \in \mathbb{C}[[t, s]]$.*

(1) *If $p = 0$ in $\mathrm{Val}^{U(n)}$, then $\frac{\partial p}{\partial t} = 0$ in $\widetilde{\mathrm{Val}}^{U(n)}$.*
(2) *If $p = 0$ in $\widetilde{\mathrm{Val}}^{U(n)}$, then $\frac{\partial(tup)}{\partial t} = 0$ in $\widetilde{\mathrm{Val}}^{U(n)}$.*
(3) *The map $p \mapsto Qp := p + \frac{\lambda}{4(1-\lambda s)} \frac{\partial(tup)}{\partial t}$ is an isomorphism of $\widetilde{\mathrm{Val}}^{U(n)}$.*

Proof A direct computation using [2, Eq. (38)] and [9, Lemma 4.4] yields

$$\frac{\partial f_{n+1}}{\partial t} = -\frac{1}{n+2}(2q_n + tq_{n-1})$$

for all n, from which the first statement follows. The second item follows from

$$\frac{\partial q_{n-1}}{\partial t} tu = (n+3)t^2 q_{n-1} + 2t q_n,$$

which can be shown by induction or by using the defining power series for q.

For the third item, we first show that $Qp := p + \frac{\lambda}{4(1-\lambda s)} \frac{\partial(put)}{t}$ is a surjective map on the space of formal power series in t, s.

Indeed, suppose that $Qp - q = \sum_{k=l}^\infty b_k(s) t^k$ for some l and some formal power series $b_k(s)$. Then

$$Q\left(p - \frac{b_l(s)}{1 + (l+1)\frac{\lambda s}{1-\lambda s}} t^l \right) - q \equiv 0 \mod t^l$$

Continuing this process with l replaced by $l + 1$ and so on, we may construct some power series p with $Qp = q$.

The space $\widetilde{\mathrm{Val}}^{U(n)}$ is a quotient of the space of formal power series in t, s, and the map Q induces a map on this quotient by (ii). This induced map is still surjective. Since $\widetilde{\mathrm{Val}}^{U(n)}$ is finite-dimensional, the map has to be an isomorphism.   □

Proposition 4.4 *Let* $\bar{w} := \bar{v} + \bar{t}\bar{u} \in \mathrm{Curv}^{U(n)*}$ *and* $p_1, p_2 \in \mathbb{C}[\bar{t}, \bar{s}]$. *Then* $p_1 + p_2\bar{w}$ *belongs to the image of* $\mathcal{V}_\lambda^n$ *if and only if*

$$\frac{\partial p_1}{\partial t} \cdot \frac{\lambda}{4(1 - \lambda s)} = p_2 \quad in \ \widetilde{\mathrm{Val}}^{U(n)}. \tag{22}$$

Proof Suppose that $p_1 + p_2\bar{w}$ belongs to the image of $\mathcal{V}_\lambda^n$. Then there is some polynomial q such that $p_1(\bar{t}, \bar{s}) + p_2(\bar{t}, \bar{s})\bar{w} = q(\bar{t}_\lambda, \bar{s})$.

By Proposition 4.2, we have

$$\bar{t}_\lambda = \frac{\bar{t}}{\sqrt{1 - \lambda\bar{s}}} + \frac{\lambda}{4(1 - \lambda\bar{s})^{\frac{3}{2}}}\bar{w}. \tag{23}$$

Since $\bar{w}^2 = 0$, the Taylor expansion of q with respect to $\bar{w}$ stops at the linear term, i.e.

$$q(\bar{t}_\lambda, \bar{s}) = q\left(\frac{\bar{t}}{\sqrt{1 - \lambda\bar{s}}}, \bar{s}\right) + q_t\left(\frac{\bar{t}}{\sqrt{1 - \lambda\bar{s}}}, \bar{s}\right)\frac{\lambda}{4(1 - \lambda\bar{s})^{\frac{3}{2}}}\bar{w},$$

where q_t denotes the partial derivative with respect to the first variable.

Using (13) and the discussion thereafter, we obtain

$$p_1(t, s) + p_2(t, s)tu = q\left(\frac{t}{\sqrt{1 - \lambda s}}, s\right) + q_t\left(\frac{t}{\sqrt{1 - \lambda s}}, s\right)\frac{\lambda}{4(1 - \lambda s)^{\frac{3}{2}}}tu \quad in \ \mathrm{Val}^{U(n)} \tag{24}$$

$$p_2(t, s) = q_t\left(\frac{t}{\sqrt{1 - \lambda s}}, s\right)\frac{\lambda}{4(1 - \lambda s)^{\frac{3}{2}}} \quad in \ \widetilde{\mathrm{Val}}^{U(n)} \tag{25}$$

Taking the derivative of the first equation with respect to t yields an equation in $\widetilde{\mathrm{Val}}^{U(n)}$ by Lemma 4.3(i). Moreover, applying Lemma 4.3(ii) to (25) will simplify this equation to

$$\frac{\partial p_1}{\partial t} = q_t\left(\frac{t}{\sqrt{1 - \lambda s}}, s\right)\frac{1}{\sqrt{1 - \lambda s}} \quad in \ \widetilde{\mathrm{Val}}^{U(n)}.$$

Multiplying by $\frac{\lambda}{4(1-\lambda s)}$ and using (25) again then yields (22).

We thus obtain that the image of $\mathcal{V}_\lambda^n$ is contained in the space of dual curvature measures satisfying (22).

Let us next compare dimensions. Rewrite $p_1 + p_2\bar{w} =: r_1 + r_2\bar{v}$ with $r_1 := p_1 + tup_2, r_2 := p_2$. Then (22) is equivalent to

$$\frac{\partial r_1}{\partial t} \cdot \frac{\lambda}{4(1 - \lambda s)} = r_2 + \frac{\lambda}{4(1 - \lambda s)}\frac{\partial(r_2 tu)}{\partial t} \quad in \ \widetilde{\mathrm{Val}}^{U(n)}. \tag{26}$$

The operator $r_2 \mapsto r_2 + \frac{\lambda}{4(1-\lambda s)} \frac{\partial (r_2 t u)}{\partial t}$ is a bijection on $\widetilde{\mathrm{Val}}^{U(n)}$ by Lemma 4.3(iii). Hence to solve (26), we can take an arbitrary $r_1 \in \mathrm{Val}^{U(n)}$, and $r_2 \in \widetilde{\mathrm{Val}}^{U(n)}$ is then uniquely determined. It follows that the dimension of the space of dual curvature measures satisfying (22) equals $\dim \mathrm{Val}^{U(n)}$, which is the same as the dimension of the image of $\mathcal{V}_\lambda^n$ in $\mathrm{Curv}^{U(n)*}$. This shows that these spaces agree.   □

5 Angular Dual Curvature Measures

Let $\mathrm{Ang}^{U(n)} \subset \mathrm{Curv}^{U(n)}$ be the subspace of angular curvature measures (see [3, Definition 2.26]). By [3, Proposition 3.2], this is the space generated by the curvature measures $\Delta_{k,q}$.

Definition 5.1 A dual curvature measure $L \in \mathrm{Curv}^{U(n)*}$ is called angular, if $L \mathrm{Ang}^{U(n)\perp} \subset \mathrm{Ang}^{U(n)\perp}$, where

$$\mathrm{Ang}^{U(n)\perp} := \{\Psi \in \mathrm{Curv}^{U(n)*} \,|\, \langle \Psi, C \rangle = 0, \forall C \in \mathrm{Ang}^{U(n)}\}$$

is the annihilator of $\mathrm{Ang}^{U(n)}$.

Lemma 5.2 *Let $p \in \mathbb{C}[t,s]$, $g \in \mathbb{C}[t,u]$ and let* PD *be the Alesker-Poincaré duality in* $\mathrm{Val}^{U(n)}$. *Then*

$$\left\langle \mathrm{PD}\left(\frac{\partial g}{\partial u}\right), psu \right\rangle = \frac{1}{8}\left\langle \mathrm{PD}(g), (-t^2 + (2n-1)u)p - 2\frac{\partial p}{\partial s}us \right\rangle.$$

Proof It suffices to check the formula for polynomials of the form $g = t^{k-2r}s^r$ (written in terms of t, u) and $p = t^{l-2i}s^i$ with $k+l = 2n-2$. Using [6, Eq. (2.57)], the left hand side equals $\omega_{2n}^{-1}\left(r\binom{2n-2(r+i+1)}{n-(r+i+1)} - \frac{r}{4}\binom{2n-2(r+i)}{n-(r+i)}\right)$, while the right hand side is $\omega_{2n}^{-1}\left(\frac{-2n+2i}{8}\binom{2n-2(r+i)}{n-(r+i)} + \frac{2n-2i-1}{2}\binom{2n-2(r+i+1)}{n-(r+i+1)}\right)$. These two expressions agree.   □

Lemma 5.3 *Set* $\bar{r}_n := \frac{4n-2}{n}\bar{t}s - \bar{t}^3$. *Then* $\bar{v} + \bar{r}_n$ *is a non-zero multiple of* $N^*_{2n-3,n-2}$.

Proof Follows from the explicit formulas (8)–(10).   □

Lemma 5.4 *A dual curvature measure L is angular if and only if* $LN^*_{2n-3,n-2} \in \mathrm{Ang}^\perp$.

Proof Since the space $\mathrm{Ang}^{U(n)}$ is spanned by the $\Delta_{k,q}$, its annihilator $\mathrm{Ang}^{U(n)\perp}$ is spanned by the $N^*_{k,q}$. Hence L is angular if and only if $LN^*_{k,q}$ is a linear combination of $N^*_{l,i}$'s. This shows the "only if" part.

By [3, Theorem 6.9] we know that $\bar{t}$ and $\bar{t}_\lambda$ are angular. Hence $\bar{t}_0' := \frac{d}{d\lambda}\big|_{\lambda=0}\bar{t}_\lambda = -\frac{1}{4}\bar{t}^3 + \frac{3}{2}\bar{t}s + \frac{1}{4}\bar{v}$ is angular.

By reverse induction on l, one can show, using Lemmas 3.2 and 3.3, that every element $N^*_{l,i}$ can be written as $q(\bar{t},\bar{t}'_0)N^*_{2n-3,n-2}$ for some polynomial q. Now suppose that $LN^*_{2n-3,n-2} \in \mathrm{Ang}^{\perp}$. Then $LN^*_{l,i} = Lq(\bar{t},\bar{t}'_0)N^*_{2n-3,n-2} = q(\bar{t},\bar{t}'_0)LN^*_{2n-3,n-2}$ is in $\mathrm{Ang}^{\perp}$ for every l, i, as claimed in the "if" part.   □

The next theorem describes the space $\mathrm{Ang}^{U(n)\perp}$.

Theorem 5.5 *Let p_1, p_2 be polynomials in t, s. Then $\bar{p}_1 + \bar{p}_2\bar{v}$ is angular if and only if*

$$tsu\left(-6tp_2 + \frac{\partial p_1}{\partial s} - \frac{\partial p_2}{\partial s}tu\right) = 0 \quad in\ \mathrm{Val}^{U(n)}.$$

Equivalently, with $\bar{w} := \bar{v} + \bar{t}\bar{u}$, $\bar{p}_1 + \bar{p}_2\bar{w}$ is angular if and only if

$$tsu\left(\frac{\partial p_1}{\partial s} - 2tp_2\right) = 0\ in\ \mathrm{Val}^{U(n)}.$$

Proof Let q_1, q_2 be polynomials in t, s. Then, by Lemma 5.3, Theorem 1.2, and Lemma 4.1 (and $\cong$ meaning equal up to some non-zero factor)

$$\langle (\bar{p}_1 + \bar{p}_2\bar{v})N^*_{2n-3,n-2}, \mathfrak{l}(q_1) + \mathfrak{n}(q_2)\rangle \cong \langle (p_1 + p_2\bar{v})(\bar{v} + r_n), \mathfrak{l}(q_1) + \mathfrak{n}(q_2)\rangle$$

$$= \langle \bar{p}_1\bar{r}_n - \bar{p}_2\bar{t}^2\bar{u}^2 + (\bar{p}_1 + \bar{p}_2\bar{r}_n - 2\bar{p}_2\bar{t}\bar{u})\bar{v}, \mathfrak{l}(q_1) + \mathfrak{n}(q_2)\rangle$$

$$= \langle \mathrm{PD}(\tilde{p}_1), q_1\rangle + \langle \mathrm{PD}(\tilde{p}_2e), q_2\rangle,$$

where $\tilde{p}_1 = p_1r_n + p_2t^2u^2$ and $\tilde{p}_2 = p_1 + p_2r_n - 2p_stu$.

By Lemma 5.4, $\bar{p}_1 + \bar{p}_2\bar{v}$ is angular if and only if $(\bar{p}_1 + \bar{p}_2\bar{v})N^*_{2n-3,n-2} \in \mathrm{Ang}^{U(n)\perp}$. By [3, Prop. 6.8.], the curvature measure $\mathfrak{l}(q_1) + \mathfrak{n}(q_2)$ is angular if and only if we can write

$$q_1 = g + 2u\frac{\partial g}{\partial u}, \quad q_2 = \frac{4t}{\pi}\frac{\partial g}{\partial u}$$

with $g \in \mathbb{R}[t, u]$.

It follows that $\bar{p}_1 + \bar{p}_2\bar{v}$ is angular if and only if

$$0 = \left\langle \mathrm{PD}(\tilde{p}_1), g + 2u\frac{\partial g}{\partial u}\right\rangle + \left\langle \mathrm{PD}(e\tilde{p}_2), \frac{4t}{\pi}\frac{\partial g}{\partial u}\right\rangle$$

$$= \langle \mathrm{PD}(\tilde{p}_1), g\rangle + \left\langle \mathrm{PD}(2u\tilde{p}_1 + \frac{4t}{\pi}e\tilde{p}_2), \frac{\partial g}{\partial u}\right\rangle$$

$$= \langle \mathrm{PD}(\tilde{p}_1), g\rangle + \left\langle \mathrm{PD}(2u(\tilde{p}_1 - tu\tilde{p}_2)), \frac{\partial g}{\partial u}\right\rangle$$

for each polynomial $g \in \mathbb{R}[t, u]$.

Plugging in the values for r_n and $\tilde{p}_1, \tilde{p}_2$, we find that

$$\tilde{p}_1 - tu\tilde{p}_2 = (p_1 - p_2 tu)(r_n - tu) = -\frac{2}{n}(p_1 - p_2 tu)ts.$$

Using Lemma 5.2 we obtain the necessary and sufficient condition for angular dual measures:

$$0 = \langle \mathrm{PD}(\tilde{p}_1), g \rangle - \frac{1}{2n}\left\langle \mathrm{PD}(t(-t^2 + (2n-1)u)(p_1 - p_2 tu) - 2t\frac{\partial(p_1 - p_2 tu)}{\partial s}us), g \right\rangle$$

$$\forall g \in \mathbb{R}[t, u].$$

By the injectivity of the Alesker-Poincaré duality, this is equivalent to

$$0 = \tilde{p}_1 - \frac{1}{2n}(t(-t^2 + (2n-1)u)(p_1 - p_2 tu) - 2t\frac{\partial(p_1 - p_2 tu)}{\partial s}us)$$

$$= \frac{tsu}{n}\left(-6tp_2 + \frac{\partial p_1}{\partial s} - \frac{\partial p_2}{\partial s}tu\right)$$

in $\mathrm{Val}^{\mathrm{U}(n)}$.   □

Proof of Theorem 1.3 Recalling (23), and $\bar{w}^2 = (\bar{v} + \bar{t}\bar{u})^2 = 0$, we obtain that

$$p(\bar{t}_\lambda, \bar{s}) = p\left(\frac{\bar{t}}{(1-\lambda\bar{s})^{\frac{1}{2}}}, \bar{s}\right) + \frac{\partial p}{\partial t}\left(\frac{\bar{t}}{(1-\lambda\bar{s})^{\frac{1}{2}}}, \bar{s}\right)\frac{\lambda}{4(1-\lambda\bar{s})^{\frac{3}{2}}}\bar{w}.$$

The statement then follows from Theorem 5.5.   □

Acknowledgements Andreas Bernig was supported by DFG grant BE 2484/5-2. Joseph H.G. Fu was supported by NSF grant DMS-1406252. Gil Solanes is a Serra Húnter Fellow and was supported by FEDER-MINECO grant MTM2015-66165-P.

References

1. S. Alesker, The multiplicative structure on continuous polynomial valuations. Geom. Funct. Anal. **14**(1), 1–26 (2004)
2. A. Bernig, J.H.G. Fu, Hermitian integral geometry. Ann. Math. **173**, 907–945 (2011)
3. A. Bernig, J.H.G. Fu, G. Solanes, Integral geometry of complex space forms. Geom. Funct. Anal. **24**(2), 403–492 (2014)
4. J.H.G. Fu, Kinematic formulas in integral geometry. Indiana Univ. Math. J. **39**(4), 1115–1154 (1990)
5. J.H.G. Fu, Structure of the unitary valuation algebra. J. Differ. Geom. **72**(3), 509–533 (2006)

6. J.H.G. Fu, Algebraic integral geometry, in *Integral Geometry and Valuations*, ed. by E. Gallego, G. Solanes. Advanced Courses in Mathematics - CRM Barcelona (Springer, Basel, 2014), pp. 47–112
7. J.H.G. Fu, D. Pokorný, J. Rataj, Kinematic formulas for sets defined by differences of convex functions. Adv. Math. **311**, 796–832 (2017)
8. T. Wannerer, Integral geometry of unitary area measures. Adv. Math. **263**, 1–44 (2014)
9. T. Wannerer, The module of unitarily invariant area measures. J. Differ. Geom. **96**(1), 141–182 (2014)

Estimates for the Integrals of Powered *i*-th Mean Curvatures

María A. Hernández Cifre and David Alonso-Gutiérrez

Abstract We show (upper and lower) estimates for the integrals of powered *i*-th mean curvatures, $i = 1, \ldots, n - 1$, of compact and convex hypersurfaces, in terms of the quermaßintegrals of the corresponding C_+^2-convex bodies. These bounds are obtained as consequences of a most general result for functions defined on a general probability space. Moreover, similar estimates for the integrals of powers of the elementary symmetric functions of the radii of curvature of C_+^2-convex bodies are proved. This probabilistic result will also allow to get new inequalities for the dual quermaßintegrals of starshaped sets, via further estimates for the integrals of the composition of a convex/concave function with the (powered) radial function.

1 Introduction

As usual in the literature we will write $\mathbb{R}^n$ for the *n*-dimensional Euclidean space, endowed with the standard inner product $\langle \cdot, \cdot \rangle$ and the Euclidean norm $\| \cdot \|$.

Moreover, $\mathscr{H}^k$, $0 \leq k \leq n$, will denote the *k*-dimensional Hausdorff measure on $\mathbb{R}^n$, and thus, if M is a subset of a *k*-plane or a *k*-dimensional sphere $\mathbb{S}^k$, then $\mathscr{H}^k(M)$ coincides, respectively, with the *k*-dimensional Lebesgue measure of M in $\mathbb{R}^k$ or with the *k*-dimensional spherical Lebesgue measure in $\mathbb{S}^k$.

A classical isoperimetric type result in differential geometry of curves due to Gage [6] states that if $\gamma : I \longrightarrow \mathbb{R}^2$ is a planar, regular, closed and convex curve with curvature k, length L and enclosing an area A, then

$$\int_\gamma k^2 \mathrm{d}s \geq \pi \frac{\mathrm{L}}{\mathrm{A}}. \tag{1}$$

M. A. Hernández Cifre (✉)
Departamento de Matemáticas, Universidad de Murcia, Murcia, Spain
e-mail: mhcifre@um.es

D. Alonso-Gutiérrez
Departamento de Matemáticas, Universidad de Zaragoza, Zaragoza, Spain
e-mail: alonsod@unizar.es

© Springer International Publishing AG 2018
G. Bianchi et al. (eds.), *Analytic Aspects of Convexity*, Springer INdAM Series 25,
https://doi.org/10.1007/978-3-319-71834-7_2

19

In [8] Green and Osher provided a general method in order to obtain inequalities of the type

$$\int_\gamma k^m \mathrm{d}s \geq f(\mathrm{L}, \mathrm{A}),$$

i.e., lower bounds for the integral of powers of the curvature in terms of some relation between the area and the length of the curve. In particular, Gage's inequality can also be derived with their method.

Moving now on to 2-dimensional surfaces in $\mathbb{R}^3$, there are two relevant curvatures to consider: the Gauss curvature κ and the mean curvature H. Then, in the spirit of (1), we find the famous Gauss-Bonnet theorem (see e.g. [5]) and the Willmore theorem (see [14]). Gauss-Bonnet's theorem shows that if $M \subset \mathbb{R}^3$ is a compact (smooth) surface which is homeomorphic to the sphere, then

$$\int_M \kappa \, \mathrm{d}\mathscr{H}^2 \geq 4\pi;$$

Willmore's inequality states that for any compact (smooth) surface $M \subset \mathbb{R}^3$ having curvature H positive everywhere,

$$\int_M H^2 \, \mathrm{d}\mathscr{H}^2 \geq 4\pi.$$

The above inequalities have their analogues for compact hypersurfaces $M \subset \mathbb{R}^n$: the Gauss-Bonnet theorem rewrites

$$\int_M \kappa \, \mathrm{d}\mathscr{H}^{n-1} \geq n|B_n|,$$

whereas Willmore's inequality becomes

$$\int_M H^{n-1} \, \mathrm{d}\mathscr{H}^{n-1} \geq n|B_n|. \tag{2}$$

Here $|\cdot|$ stands for the volume, i.e., the Lebesgue measure, and B_n denotes the Euclidean unit ball centered at the origin. Willmore's inequality in an arbitrary dimension was proved by Chen, see [3, 4]. In addition, Ros [11] proved that

$$\int_M \frac{1}{H} \, \mathrm{d}\mathscr{H}^{n-1} \geq n|M|. \tag{3}$$

Besides the major importance that these results have by themselves, they are specially interesting because they imply isoperimetric inequalities (see e.g. [10]).

Since a compact hypersurface $M \subset \mathbb{R}^n$ has associated $n-1$ relevant curvatures, the so-called i-th mean curvatures H_i, $i = 1, \ldots, n-1$, the above results motivate the following question:

Main problem: To obtain (lower and/or upper) estimates for the integrals

$$\int_M H_i^\alpha \, d\mathcal{H}^{n-1} \quad \text{and} \quad \int_M \frac{1}{H_i^\alpha} \, d\mathcal{H}^{n-1}$$

for $\alpha \geq 0$ and any $i = 1, \ldots, n-1$, as well as improvements of Chen's and Ros' inequalities, in the convex case.

Next we introduce the notation and main concepts that will be needed throughout the paper, as well as our main results.

2 Notation and Previous Results

Let $\mathcal{K}_0^n$ be the set of all convex bodies, i.e., compact convex sets with non-empty interior, in $\mathbb{R}^n$ containing the origin 0. A convex body $K \in \mathcal{K}_0^n$ is said to be of class C^2 if its boundary hypersurface bd K is a regular submanifold of $\mathbb{R}^n$, in the sense of differential geometry, which is twice continuously differentiable. Moreover, we say that K is of class C_+^2 if K is of class C^2 and the Gauss map $\nu_K : \text{bd } K \longrightarrow \mathbb{S}^{n-1}$, mapping a boundary point $x \in \text{bd } K$ to the (unique) normal vector of K at x, is a diffeomorphism. Thus, in this case, we can consider the $n-1$ principal curvatures $k_1, \ldots, k_{n-1}$ of bd K and, as usual in the literature, we will denote by

$$H_i = \frac{1}{\binom{n-1}{i}} \sum_{1 \leq j_1 < \cdots < j_i \leq n-1} k_{j_1} \cdots k_{j_i}, \quad i = 1, \ldots, n-1,$$

the i-th mean curvature, setting $H_0 = 1$. In particular, $H_1 = H$ is the classical *mean curvature* and $H_{n-1} = \kappa$ is the *Gauss-Kronecker curvature*.

The pursued estimates for the integrals of powered i-th mean curvatures will be given in terms of the so-called quermaßintegrals of the convex bodies, which are special geometric measures associated to the set. We define them next.

In [13], Steiner proved that given $K \in \mathcal{K}_0^n$ and a non-negative real number λ, the volume of the Minkowski sum (vectorial addition) $K + \lambda B_n$ is expressed as a polynomial of degree n in λ, namely,

$$|K + \lambda B_n| = \sum_{i=0}^{n} \binom{n}{i} W_i(K) \lambda^i,$$

which is called the (classical) *Steiner formula* of K. The coefficients $W_i(K)$ are the *quermaßintegrals* of K, and they are a special case of the more general defined *mixed volumes* for which we refer to [12, Section 5.1]. In particular, $W_0(K) = |K|$ (the area $A(K)$ in the planar case), $nW_1(K) = S(K)$ is the surface area (the perimeter $L(K)$ in the plane) and $2W_{n-1}/|B_n|$ is the mean width of K. Moreover, $W_n(K) = |B_n|$.

In order to state the main results of the paper, we need an additional definition. The *outer radius* and the *inner radius* of $K \in \mathscr{K}_0^n$ are defined as the quantities

$$\bar{R}(K) = \min\{R > 0 : K \subseteq RB_n\} = \max\{\|x\| : x \in \mathrm{bd}\, K\},$$

$$\bar{r}(K) = \max\{r \geq 0 : rB_n \subseteq K\} = \min\{\|x\| : x \in \mathrm{bd}\, K\}.$$

Clearly, the value $\bar{R}(K) - \bar{r}(K)$ is not translation invariant, but since K is compact, there exists a (unique) point $c_K \in K$ such that

$$\bar{R}(K - c_K) - \bar{r}(K - c_K) = \min\{\bar{R}(K - t) - \bar{r}(K - t) : t \in K\}$$

(see [2]). The point c_K is the *center of the minimal ring*, i.e., the uniquely determined ring (closed set consisting of all points between two concentric balls) with minimal difference of radii containing bd K. The value $\omega_a(K) := \bar{R}(K - c_K) - \bar{r}(K - c_K)$ is called the *width of the minimal ring* of K (Fig. 1).

Since $\bar{r}B_n \subseteq K$ and $K \subseteq \bar{R}B_n$, the in- and outer radii and the quermaßintegrals relate in the following way:

$$\bar{r}(K)W_{i+1}(K) \leq W_i(K) \leq \bar{R}(K)W_{i+1}(K), \tag{4}$$

$i = 0, \ldots, n - 1$, which is a direct consequence of the monotonicity of the mixed volumes (cf. e.g. [12, p. 282]).

For the statement of the results, and in order to shorten the statements and proofs, we introduce the following notation. For $1 \leq j \leq n - 1$ and $0 \leq k \leq n - 1$, let

$$\eta_{j,k} = \begin{cases} \dfrac{W_1(K)W_j(K) - W_0(K)W_{j+1}(K)}{W_{k+1}(K)^2 \omega_a(K)} & \text{if } K \neq rB_n \text{ for all } r > 0, \\ 0 & \text{if } K = rB_n. \end{cases}$$

Fig. 1 The minimal ring of a convex body. We observe that the inner and the outer radii of $K - c_K$ do not necessarily coincide with the classical inradius and circumradius of the set, respectively

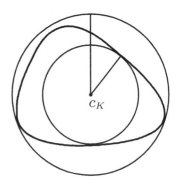

The non-negativity of the values $\eta_{j,k}$ is a direct consequence of the inequalities

$$W_i(K)W_j(K) \geq W_{i-1}(K)W_{j+1}(K), \quad 1 \leq i \leq j \leq n-1, \tag{5}$$

particular cases of the Aleksandrov-Fenchel inequality (see e.g. [12, Section 7.3]). From now on, for the sake of brevity, we will write $W_i = W_i(K)$, $i = 0, \dots, n$, and analogously for all other functionals, if the distinction of the body is not needed.

2.1   Some Previous Results

In [1] the above mentioned problem of obtaining lower estimates for the integral of powered *i*-th mean curvatures was studied. Among others, the following more general theorem was proved.

Theorem 1 *Let $K \in \mathcal{K}_0^n$ be of class C_+^2. Then, for any convex function $F : I \longrightarrow \mathbb{R}$, $I \subseteq \mathbb{R}$ where all the quantities are defined, and all $i = 0, \dots, n-1$,*

$$\int_{\mathrm{bd}\,K} (F \circ H_i)\, \mathrm{d}\mathcal{H}^{n-1} \geq n W_1 \frac{F\left(\frac{W_{i+1}}{W_1} + \eta_{i,0}\right) + F\left(\frac{W_{i+1}}{W_1} - \eta_{i,0}\right)}{2},$$

$$\int_{\mathrm{bd}\,K} \left(F \circ \frac{1}{H_i}\right) H_i\, \mathrm{d}\mathcal{H}^{n-1} \geq n W_{i+1} \frac{F\left(\frac{W_1}{W_{i+1}} + \eta_{i,i}\right) + F\left(\frac{W_1}{W_{i+1}} - \eta_{i,i}\right)}{2}.$$

Equality holds in both inequalities if $K = B_n$ (up to dilations).

Indeed, a slightly more general result was obtained (see [1, Theorem 3.2]).

Then, applying Theorem 1 to the convex functions $F(x) = x^{\alpha+1}$ or $F(x) = 1/x^{\alpha}$, $\alpha \geq 0$, two different results can be obtained, providing different bounds for the same integrals. These bounds can be compared, and thus the following theorem is obtained in the spirit of the **main problem**:

Theorem 2 *Let $K \in \mathcal{K}_0^n$ be of class C_+^2. Then, for any $\alpha \geq 0$ and all $i = 0, \dots, n-1$,*

$$\int_{\mathrm{bd}\,K} H_i^{\alpha+1}\, \mathrm{d}\mathcal{H}^{n-1} \geq \frac{n}{2} \left[\frac{W_{i+1}^{\alpha+1}}{(W_1 + W_{i+1}\eta_{i,i})^{\alpha}} + \frac{W_{i+1}^{\alpha+1}}{(W_1 - W_{i+1}\eta_{i,i})^{\alpha}} \right],$$

$$\int_{\mathrm{bd}\,K} \frac{1}{H_i^{\alpha}}\, \mathrm{d}\mathcal{H}^{n-1} \geq \frac{n}{2} \left[\frac{W_1^{\alpha+1}}{(W_{i+1} + W_1\eta_{i,0})^{\alpha}} + \frac{W_1^{\alpha+1}}{(W_{i+1} - W_1\eta_{i,0})^{\alpha}} \right].$$

Equality holds in both inequalities if $K = B_n$ (up to dilations).

In particular, improvements of Chen's and Ros' estimates for convex hypersurfaces can be obtained by just taking $i = 1$ and, respectively, $\alpha = 1$ or $\alpha = n - 2$:

Corollary 1 *Let* $K \in \mathcal{K}_0^n$ *be of class* C_+^2. *Then,*

$$\int_{\mathrm{bd}\,K} H^{n-1} \mathrm{d}\mathcal{H}^{n-1} \geq \frac{n}{2} \left[\frac{\mathrm{W}_2^{n-1}}{\left(\mathrm{W}_1 + \mathrm{W}_2 \eta_{1,1}\right)^{n-2}} + \frac{\mathrm{W}_2^{n-1}}{\left(\mathrm{W}_1 - \mathrm{W}_2 \eta_{1,1}\right)^{n-2}} \right],$$

$$\int_{\mathrm{bd}\,K} \frac{1}{H} \mathrm{d}\mathcal{H}^{n-1} \geq n \frac{\mathrm{W}_1^2 \mathrm{W}_2}{\mathrm{W}_2^2 - \mathrm{W}_1^2 \eta_{1,0}^2}.$$

Equality holds in all inequalities if $K = B_n$ *(up to dilations).*

Indeed, on the one hand,

$$n \frac{\mathrm{W}_1^2 \mathrm{W}_2}{\mathrm{W}_2^2 - \mathrm{W}_1^2 \eta_{1,0}^2} \geq n \frac{\mathrm{W}_1^2 \mathrm{W}_2}{\mathrm{W}_2^2} = n \frac{\mathrm{W}_1^2}{\mathrm{W}_2} \geq n \mathrm{W}_0 = n|K|$$

because of the Aleksandrov-Fenchel inequality (5) for $i = j = 1$. On the other hand, since the function $1/x^{n-2}$ is convex, then

$$\frac{n}{2} \left[\frac{\mathrm{W}_2^{n-1}}{\left(\mathrm{W}_1 + \mathrm{W}_2 \eta_{1,1}\right)^{n-2}} + \frac{\mathrm{W}_2^{n-1}}{\left(\mathrm{W}_1 - \mathrm{W}_2 \eta_{1,1}\right)^{n-2}} \right] \geq \frac{n}{2} \frac{2 \mathrm{W}_2^{n-1}}{\mathrm{W}_1^{n-2}} \geq n|B_n|,$$

where the last inequality follows from the known relations

$$\mathrm{W}_j^{k-i} \geq \mathrm{W}_i^{k-j} \mathrm{W}_k^{j-i} \quad \text{for } 0 \leq i < j < k \leq n, \tag{6}$$

which are also consequences of the Aleksandrov-Fenchel inequality (see e.g. [12, (7.66)]). Hence, Corollary 1 improves (2) and (3) in the convex case.

We notice that Theorem 2 provides lower estimates for the integral of *almost any* power of the i-th mean curvatures: bounds for $\int_{\mathrm{bd}\,K} H_i^\lambda \mathrm{d}\mathcal{H}^{n-1}$ are given *for any* $\lambda \in (-\infty, 0] \cup [1, +\infty)$; the range $(0, 1)$ is still an open question.

In this work we consider the opposite case, i.e., we will look for upper bounds for the integrals of powered i-th mean curvatures.

3   A Probabilistic Type Result

All the results will be consequences of a very general proposition for functions defined on a general probability space. In [1, Proposition 1.4], this result was proved for a convex function F, whereas now we are interested in the concave case. Although this one can be obtained from the convex case just taking $-F$, for completeness, we include here the proof.

As usual in the literature,

$$\mathbb{E}\rho = \int_\Omega \rho(\omega)\,d\mathbb{P}(\omega)$$

will denote the expectation of ρ, $\mathrm{Cov}(\rho, h) = \mathbb{E}h\rho - \mathbb{E}h\mathbb{E}\rho$ the covariance of ρ and h, and $\|\cdot\|_\infty$ the sup-norm, i.e., $\|f\|_\infty = \sup\{|f(\omega)| : \omega \in \Omega\}$.

Proposition 1 *Let $(\Omega, \mathbb{P})$ be a probability space such that, for any $A \subseteq \Omega$ and any $0 \le p \le \mathbb{P}(A)$, there exists $B \subseteq A$ with $\mathbb{P}(B) = p$. Let $\rho, h : \Omega \longrightarrow \mathbb{R}$, with $\rho \in L^1(\Omega)$ and $h \in L^\infty(\Omega)$. Then, for any concave function $F : I \longrightarrow \mathbb{R}$, $I \subseteq \mathbb{R}$ where all the expressions below are defined, we have*

$$\mathbb{E}(F \circ \rho) \le \frac{F\left(\mathbb{E}\rho + \frac{\mathrm{Cov}(\rho, h)}{\|h - \mathbb{E}h\|_\infty}\right) + F\left(\mathbb{E}\rho - \frac{\mathrm{Cov}(\rho, h)}{\|h - \mathbb{E}h\|_\infty}\right)}{2}.$$

Proof Without loss of generality we assume that $\mathrm{Cov}(\rho, h) \le 0$; otherwise we just change h by $-h$.

Let m be a median of ρ, i.e., a value for which both

$$\mathbb{P}(\{\omega \in \Omega : \rho(\omega) \ge m\}) \ge 1/2 \quad \text{and} \quad \mathbb{P}(\{\omega \in \Omega : \rho(\omega) \le m\}) \ge 1/2,$$

and let $\Omega_1 \subset \Omega$ and $\Omega_2 = \Omega \setminus \Omega_1$ be such that $\mathbb{P}(\Omega_1) = \mathbb{P}(\Omega_2) = 1/2$ and

$$\{\omega \in \Omega : \rho(\omega) > m\} \subseteq \Omega_1 \subseteq \{\omega \in \Omega : \rho(\omega) \ge m\},$$
$$\{\omega \in \Omega : \rho(\omega) < m\} \subseteq \Omega_2 \subseteq \{\omega \in \Omega : \rho(\omega) \le m\}.$$

We notice that such Ω_1 always exists. Indeed, by the definition of median,

$$\mathbb{P}\left(\{\omega \in \Omega : \rho(\omega) \le m\}\right) \ge \frac{1}{2},$$

and so

$$\mathbb{P}\left(\{\omega \in \Omega : \rho(\omega) > m\}\right) \le \frac{1}{2}.$$

Consequently, since

$$\mathbb{P}\left(\{\omega \in \Omega : \rho(\omega) \ge m\}\right)$$

$$= \mathbb{P}\left(\{\omega \in \Omega : \rho(\omega) > m\}\right) + \mathbb{P}\left(\{\omega \in \Omega : \rho(\omega) = m\}\right) \ge \frac{1}{2},$$

we have that

$$\mathbb{P}\Big(\big\{\omega \in \Omega : \rho(\omega) = m\big\}\Big) \geq \frac{1}{2} - \mathbb{P}\Big(\big\{\omega \in \Omega : \rho(\omega) > m\big\}\Big) \geq 0.$$

Then, by our assumptions on $(\Omega, \mathbb{P})$, there exists a subset $B \subseteq \{\omega \in \Omega : \rho(\omega) = m\}$ with $\mathbb{P}(B) = 1/2 - \mathbb{P}\{\omega \in \Omega : \rho(\omega) > m\}$ and we can take

$$\Omega_1 = \big\{\omega \in \Omega : \rho(\omega) > m\big\} \cup B.$$

Now, let

$$\rho_1 = 2 \int_{\Omega_1} \rho(\omega)\, d\mathbb{P}(\omega) \quad \text{and} \quad \rho_2 = 2 \int_{\Omega_2} \rho(\omega)\, d\mathbb{P}(\omega).$$

Since $\rho_1 + \rho_2 = 2\,\mathbb{E}\rho$, we can write

$$\rho_1 = \mathbb{E}\rho + b \quad \text{and} \quad \rho_2 = \mathbb{E}\rho - b \tag{7}$$

for some $b \geq 0$. First, we are going to prove that

$$\frac{\big|\operatorname{Cov}(\rho, h)\big|}{\|h - \mathbb{E}h\|_\infty} \leq b. \tag{8}$$

Indeed, since

$$-\|h - \mathbb{E}h\|_\infty \leq \mathbb{E}h - h(\omega) \leq \|h - \mathbb{E}h\|_\infty$$

for every $\omega \in \Omega$ and since $\rho(\omega) \geq m$ if $\omega \in \Omega_1$ and $\rho(\omega) \leq m$ if $\omega \in \Omega_2$, then we have that

$$\int_{\Omega_1} \big(\mathbb{E}h - h(\omega)\big)\big(\rho(\omega) - m\big)\, d\mathbb{P}(\omega) \leq \frac{1}{2}\|h - \mathbb{E}h\|_\infty(\rho_1 - m)$$

and

$$\int_{\Omega_2} \big(\mathbb{E}h - h(\omega)\big)\big(\rho(\omega) - m\big)\, d\mathbb{P}(\omega) \leq -\frac{1}{2}\|h - \mathbb{E}h\|_\infty(\rho_2 - m).$$

Adding both integrals and using (7) we get

$$\mathbb{E}\big((\mathbb{E}h - h)(\rho - m)\big) = \int_{\Omega} \big(\mathbb{E}h - h(\omega)\big)\big(\rho(\omega) - m\big)\, d\mathbb{P}(\omega)$$

$$\leq \frac{1}{2}\|h - \mathbb{E}h\|_\infty(\rho_1 - \rho_2) = \|h - \mathbb{E}h\|_\infty b,$$

and since

$$\mathbb{E}\big((\mathbb{E}h - h)(\rho - m)\big) = \mathbb{E}h\mathbb{E}\rho - \mathbb{E}h\rho = -\operatorname{Cov}(\rho, h),$$

we obtain the required bound (8).

Now, since F is concave, Jensen's inequality (see e.g. [12, p. 20]) yields

$$F(\rho_1) \geq 2 \int_{\Omega_1} (F \circ \rho)(\omega)\, d\mathbb{P}(\omega) \quad \text{and} \quad F(\rho_2) \geq 2 \int_{\Omega_2} (F \circ \rho)(\omega)\, d\mathbb{P}(\omega),$$

which, together with (7) implies that

$$\begin{aligned}
\mathbb{E}(F \circ \rho) &= \int_{\Omega_1} (F \circ \rho)(\omega)\, d\mathbb{P}(\omega) + \int_{\Omega_2} (F \circ \rho)(\omega)\, d\mathbb{P}(\omega) \\
&\leq \frac{F(\rho_1) + F(\rho_2)}{2} = \frac{F(\mathbb{E}\rho + b) + F(\mathbb{E}\rho - b)}{2}.
\end{aligned}$$

Finally, since a concave function F satisfies that for any $x \in \mathbb{R}$ and any $0 \leq a \leq b$ the average of the numbers $\{F(x + a), F(x - a)\}$ is not smaller than the average of $\{F(x + b), F(x - b)\}$, taking into account (8) we get

$$\mathbb{E}(F \circ \rho) \leq \frac{F(\mathbb{E}\rho + b) + F(\mathbb{E}\rho - b)}{2} \leq \frac{F\left(\mathbb{E}\rho + \frac{\operatorname{Cov}(\rho, h)}{\|h - \mathbb{E}h\|_\infty}\right) + F\left(\mathbb{E}\rho - \frac{\operatorname{Cov}(\rho, h)}{\|h - \mathbb{E}h\|_\infty}\right)}{2},$$

which concludes the proof. $\qquad\qquad\qquad\qquad\qquad\qquad\qquad\qquad\qquad\qquad\qquad\square$

If the probability measure can be expressed by means of a density with respect to another (not necessarily a probability) measure μ, we immediately obtain the following result.

Proposition 2 *Let (Ω, μ) be a measure space and let $g : \Omega \longrightarrow \mathbb{R}$ be a positive integrable function with $\int_\Omega g\, d\mu = 1$, and such that for any $A \subseteq \Omega$ and any $0 \leq p \leq \int_A g\, d\mu$, there exists $B \subseteq A$ with $\int_B g\, d\mu = p$. Let $\rho, h : \Omega \longrightarrow \mathbb{R}$ be integrable functions with $h \in L^\infty(\Omega)$. Then, for any concave function $F : I \longrightarrow \mathbb{R}, I \subseteq \mathbb{R}$ where all the expressions below are defined, we have*

$$\int_\Omega (F \circ \rho) g\, d\mu \leq \frac{F\left(\int_\Omega \rho g\, d\mu + \eta(\rho, h, g)\right) + F\left(\int_\Omega \rho g\, d\mu - \eta(\rho, h, g)\right)}{2},$$

where

$$\eta(\rho, h, g) = \frac{\int_\Omega \rho h g\, d\mu - \left(\int_\Omega \rho g\, d\mu\right)\left(\int_\Omega h g\, d\mu\right)}{\left\|h - \int_\Omega h g\, d\mu\right\|_\infty}.$$

4 Upper Bounds for Integrals of Powered i-th Mean Curvatures

We denote by $h_K(u) = \sup_{x \in K} \langle x, u \rangle$, $u \in \mathbb{R}^n$, the support function of K (see e.g. [12, Section 1.7]), and let

$$q_K(x) = h_K(v_K(x)) = \langle x, v_K(x) \rangle, \quad x \in \mathrm{bd}\, K.$$

Minkowskian integral formulae (see e.g. [12, pp. 296–297]) state that

$$\mathrm{W}_i = \frac{1}{n} \int_{\mathrm{bd}\, K} H_{i-1} \, \mathrm{d}\mathcal{H}^{n-1} = \frac{1}{n} \int_{\mathrm{bd}\, K} q_K H_i \, \mathrm{d}\mathcal{H}^{n-1} \tag{9}$$

for $i = 1, \ldots, n$. We observe that the volume

$$|K| = \mathrm{W}_0 = \frac{1}{n} \int_{\mathrm{bd}\, K} q_K H_0 \, \mathrm{d}\mathcal{H}^{n-1} = \frac{1}{n} \int_{\mathrm{bd}\, K} q_K \, \mathrm{d}\mathcal{H}^{n-1}.$$

This section is devoted to look for upper bounds for the integrals of some powers of the i-th mean curvatures. First we show the following general result for an arbitrary concave function.

Theorem 3 *Let $K \in \mathcal{K}_0^n$ be of class C_+^2. For any concave function $F : I \longrightarrow \mathbb{R}$, $I \subseteq \mathbb{R}$ where all the quantities are defined, and all $i = 0, \ldots, n-1$,*

$$\int_{\mathrm{bd}\, K} (F \circ H_i) \, \mathrm{d}\mathcal{H}^{n-1} \le n\mathrm{W}_1 \frac{F\left(\frac{\mathrm{W}_{i+1}}{\mathrm{W}_1} + \eta_{i,0}\right) + F\left(\frac{\mathrm{W}_{i+1}}{\mathrm{W}_1} - \eta_{i,0}\right)}{2}, \tag{10}$$

$$\int_{\mathrm{bd}\, K} \left(F \circ \frac{1}{H_i}\right) H_i \, \mathrm{d}\mathcal{H}^{n-1} \le n\mathrm{W}_{i+1} \frac{F\left(\frac{\mathrm{W}_1}{\mathrm{W}_{i+1}} + \eta_{i,i}\right) + F\left(\frac{\mathrm{W}_1}{\mathrm{W}_{i+1}} - \eta_{i,i}\right)}{2}. \tag{11}$$

Equality holds in both inequalities if $K = B_n$ (up to dilations).

Proof In order to get (10), we consider the probability space $\left(\mathrm{bd}\, K, \mathcal{H}^{n-1}/(n\mathrm{W}_1)\right)$ and apply Proposition 1 to the functions $\rho = H_i$ and $h = q_K$. Then, using the identities in (9), we get

$$\mathbb{E}\rho = \frac{\mathrm{W}_{i+1}}{\mathrm{W}_1}, \qquad \mathbb{E}h = \frac{\mathrm{W}_0}{\mathrm{W}_1}$$

and

$$\mathrm{Cov}(\rho, h) = \mathbb{E}h\rho - \mathbb{E}h\mathbb{E}\rho = \frac{\mathrm{W}_i\mathrm{W}_1 - \mathrm{W}_0\mathrm{W}_{i+1}}{\mathrm{W}_1^2}.$$

Moreover,

$$\|h - \mathbb{E}h\|_\infty = \sup\left\{\left|q_K(x) - \frac{W_0}{W_1}\right| : x \in \operatorname{bd} K\right\} = \max\left\{\bar{R} - \frac{W_0}{W_1}, \frac{W_0}{W_1} - \bar{r}\right\},$$

and since the functionals H_j, W_j are translation invariant, the smallest possible upper bound for $\int_{\operatorname{bd} K}(F \circ H_i)\, d\mathcal{H}^{n-1}$ will be obtained for the translation of K such that the above maximum is minimal. Therefore, we can write

$$\int_{\operatorname{bd} K}(F \circ H_i)\, d\mathcal{H}^{n-1} \le nW_1 \frac{F\left(\frac{W_{i+1}}{W_1} + \eta\right) + F\left(\frac{W_{i+1}}{W_1} - \eta\right)}{2},$$

with

$$\delta = \frac{\operatorname{Cov}(\rho, h)}{\|h - \mathbb{E}h\|_\infty} = \frac{W_iW_1 - W_0W_{i+1}}{W_1^2 \min_{x\in K}\max\left\{\bar{R}(K - x) - \frac{W_0}{W_1}, \frac{W_0}{W_1} - \bar{r}(K - x)\right\}}.$$

Now we observe that, by (4),

$$\bar{R}(K - x) - \frac{W_0}{W_1} \le \bar{R}(K - x) - \bar{r}(K - x), \qquad \text{and}$$

$$\frac{W_0}{W_1} - \bar{r}(K - x) \le \bar{R}(K - x) - \bar{r}(K - x),$$

and since F is a concave function, we can replace δ by a smaller number, namely,

$$\delta \ge \frac{W_iW_1 - W_0W_{i+1}}{W_1^2 \min_{x\in K}\left\{\bar{R}(K - x) - \bar{r}(K - x)\right\}} = \frac{W_iW_1 - W_0W_{i+1}}{W_1^2 \omega_a} = \eta_{i,0}.$$

Altogether we get (10).

Inequality (11) is obtained analogously, but now as a consequence of Proposition 2 for $\rho = 1/H_i$, $h = q_K$ and $g = H_i/(nW_{i+1})$; we notice that, by (9), $\int_{\operatorname{bd} K} g\, d\mathcal{H}^{n-1} = 1$.

Finally, equality trivially holds for $K = B_n$ (up to dilations) just noticing that $W_i(B_n) = |B_n|$ for all $i = 0, \ldots, n$. $\qquad\square$

In order to get bounds for the integral of some powers of the mean curvatures, we may apply Theorem 3 to the concave function $F(x) = x^\alpha$ for $0 \le \alpha \le 1$.

Theorem 4 *Let $K \in \mathcal{K}_0^n$ be of class C_+^2. Then, for all $i = 0, \ldots, n-1$, the following inequalities hold:*

- *If $0 \le \alpha \le 1/2$,*

$$\int_{\operatorname{bd} K} H_i^\alpha\, d\mathcal{H}^{n-1} \le \frac{n}{2}W_1^{1-\alpha}\left[\left(W_{i+1} + W_1\, \eta_{i,0}\right)^\alpha + \left(W_{i+1} - W_1\, \eta_{i,0}\right)^\alpha\right]. \quad (12)$$

- If $1/2 \leq \alpha \leq 1$,

$$\int_{\mathrm{bd}\,K} H_i^\alpha \, \mathrm{d}\mathcal{H}^{n-1} \leq \frac{n}{2} \mathrm{W}_{i+1}^\alpha \left[\left(\mathrm{W}_1 + \mathrm{W}_{i+1}\, \eta_{i,i} \right)^{1-\alpha} + \left(\mathrm{W}_1 - \mathrm{W}_{i+1}\, \eta_{i,i} \right)^{1-\alpha} \right]. \qquad (13)$$

Equality holds in both inequalities if $K = B_n$ (up to dilations).

Proof Let $0 \leq \alpha \leq 1$. On the one hand, taking $F(x) = x^\alpha$ in (10) we directly get

$$\int_{\mathrm{bd}\,K} H_i^\alpha \, \mathrm{d}\mathcal{H}^{n-1} \leq \frac{n}{2} \mathrm{W}_1^{1-\alpha} \left[\left(\mathrm{W}_{i+1} + \mathrm{W}_1\, \eta_{i,0} \right)^\alpha + \left(\mathrm{W}_{i+1} - \mathrm{W}_1\, \eta_{i,0} \right)^\alpha \right]. \qquad (14)$$

On the other hand, (11) applied to $F(x) = x^\alpha$ yields

$$\int_{\mathrm{bd}\,K} H_i^{1-\alpha} \, \mathrm{d}\mathcal{H}^{n-1} \leq \frac{n}{2} \mathrm{W}_{i+1}^{1-\alpha} \left[\left(\mathrm{W}_1 + \mathrm{W}_{i+1}\, \eta_{i,i} \right)^\alpha + \left(\mathrm{W}_1 - \mathrm{W}_{i+1}\, \eta_{i,i} \right)^\alpha \right]$$

or, equivalently,

$$\int_{\mathrm{bd}\,K} H_i^\alpha \, \mathrm{d}\mathcal{H}^{n-1} \leq \frac{n}{2} \mathrm{W}_{i+1}^\alpha \left[\left(\mathrm{W}_1 + \mathrm{W}_{i+1}\, \eta_{i,i} \right)^{1-\alpha} + \left(\mathrm{W}_1 - \mathrm{W}_{i+1}\, \eta_{i,i} \right)^{1-\alpha} \right]. \qquad (15)$$

Therefore, we just have to compare both bounds, depending on the value of α. In order to do it, we denote by

$$x = \frac{\mathrm{W}_1}{\mathrm{W}_{i+1}} \eta_{i,0} = \frac{\mathrm{W}_{i+1}}{\mathrm{W}_1} \eta_{i,i} = \frac{\mathrm{W}_1 \mathrm{W}_i - \mathrm{W}_0 \mathrm{W}_{i+1}}{\mathrm{W}_1 \mathrm{W}_{i+1} \omega_a}.$$

Using (4) and (5) we get that

$$0 \leq \mathrm{W}_1 \mathrm{W}_i - \mathrm{W}_0 \mathrm{W}_{i+1} \leq \mathrm{W}_1 \mathrm{W}_{i+1} \left(\bar{\mathrm{R}}(K - c_K) - \bar{\mathrm{r}}(K - c_K) \right) = \mathrm{W}_1 \mathrm{W}_{i+1} \omega_a,$$

and therefore, $0 \leq x \leq 1$. Using this notation, the upper bounds in (14) and (15) can be written, respectively, as

$$\mathrm{W}_1^{1-\alpha} \mathrm{W}_{i+1}^\alpha \left[(1 + x)^\alpha + (1 - x)^\alpha \right] =: (\mathrm{b}1),$$

$$\mathrm{W}_1^{1-\alpha} \mathrm{W}_{i+1}^\alpha \left[(1 + x)^{1-\alpha} + (1 - x)^{1-\alpha} \right] =: (\mathrm{b}2).$$

Then (b1) is, say, smaller than (b2), if and only if

$$(1 + x)^\alpha + (1 - x)^\alpha \leq (1 + x)^{1-\alpha} + (1 - x)^{1-\alpha}, \qquad (16)$$

and it clearly holds when $\alpha \leq 1/2$. This shows (12). Finally, (b1) $\geq$ (b2) is equivalent to have the reverse inequality in (16), which holds if $\alpha \geq 1/2$. It states (13) and concludes the proof of the theorem. $\square$

It may also have interest to obtain an estimate for the entropy of the *i*-th mean curvatures, which is defined by

$$-\int_{\mathrm{bd}\, K} H_i \log H_i \, \mathrm{d}\mathscr{H}^{n-1}.$$

We do it in the following result.

Corollary 2 *Let $K \in \mathscr{K}_0^n$ be of class C_+^2. For all $i = 0, \ldots, n-1$,*

$$-\int_{\mathrm{bd}\, K} H_i \log H_i \, \mathrm{d}\mathscr{H}^{n-1} \leq \frac{n}{2} W_{i+1} \log\left(\frac{W_1^2}{W_{i+1}^2} - \eta_{i,i}^2\right).$$

Equality holds if $K = B_n$ (up to dilations).

Proof It is a direct consequence of inequality (11), just considering the concave function $F(x) = \log x$:

$$-\int_{\mathrm{bd}\, K} H_i \log H_i \, \mathrm{d}\mathscr{H}^{n-1} = \int_{\mathrm{bd}\, K} H_i \log \frac{1}{H_i} \, \mathrm{d}\mathscr{H}^{n-1} \leq \frac{n}{2} W_{i+1} \log\left(\frac{W_1^2}{W_{i+1}^2} - \eta_{i,i}^2\right).$$

$\square$

At this point we would like to mention that it might be interesting to study this problem in the non-smooth case, working, for instance, with the generalized principal curvatures, which can be defined for any convex body. Then the corresponding elementary symmetric functions of those generalized curvatures should be considered. In the case of sets of positive reach, the curvature measures (of Federer) have integral expressions, over the normal bundle and with respect to the Hausdorff measure, of these elementary symmetric functions (see e.g. [12, Notes for s. 4.2] and the references therein). Therefore, it can be studied whether our approach can be useful in this setting.

4.1   On the Radii of Curvature of Convex Bodies

If $K \in \mathscr{K}_0^n$ is of class C_+^2, we can consider the $n-1$ principal radii of curvature $r_1, \ldots, r_{n-1}$ of K at $u \in \mathbb{S}^{n-1}$, i.e., the eigenvalues of the reverse Weingarten map (see e.g. [12, p. 116] for a detailed explanation). Then, for $i = 1, \ldots, n-1$,

$$s_i = \frac{1}{\binom{n-1}{i}} \sum_{1 \leq j_1 < \cdots < j_i \leq n-1} r_{j_1} \cdots r_{j_i}$$

is the i-th normalized elementary symmetric function of the principal radii of curvature, with $s_0 = 1$. We observe that, properly ordering the indices,

$$r_i(u) = \frac{1}{k_i(x_K(u))}, \quad i = 1, \ldots, n-1,$$

where $x_K(u) \in \mathrm{bd}\, K$ is the unique point of the boundary at which u is the outer normal vector. Moreover, for all $u \in \mathbb{S}^{n-1}$ and $x \in \mathrm{bd}\, K$ we have the relations

$$s_i(u) = \frac{H_{n-i-1}}{H_{n-1}}(x_K(u)) \quad \text{and} \quad H_i(x) = \frac{s_{n-i-1}}{s_{n-1}}(v_K(x)),$$

and so there exist also Minkowskian integral formulae for the s_i's (see e.g. [12, pp. 296–297]): for all $i = 0, \ldots, n-1$,

$$W_i = \frac{1}{n} \int_{\mathbb{S}^{n-1}} s_{n-i} \, \mathrm{d}\mathcal{H}^{n-1} = \frac{1}{n} \int_{\mathbb{S}^{n-1}} h_K s_{n-i-1} \, \mathrm{d}\mathcal{H}^{n-1}. \tag{17}$$

In [1] the analogous result to Theorem 1 for the so-called i-th (normalized) elementary symmetric function of the principal radii of curvature was obtained. In a similar way we can also get the corresponding result for a concave function, which will provide upper bounds for the integrals of some powers of the s_i's.

Theorem 5 *Let $K \in \mathcal{K}_0^n$ be of class C_+^2. For any concave function $F : I \longrightarrow \mathbb{R}$, $I \subseteq \mathbb{R}$ where all the quantities are defined, and all $i = 0, \ldots, n-1$,*

$$\int_{\mathbb{S}^{n-1}} (F \circ s_i) \, \mathrm{d}\mathcal{H}^{n-1} \leq n|B_n| \frac{F\left(\frac{W_{n-i}}{|B_n|} + \bar{\eta}_{i,1}\right) + F\left(\frac{W_{n-i}}{|B_n|} - \bar{\eta}_{i,1}\right)}{2},$$

$$\int_{\mathbb{S}^{n-1}} \left(F \circ \frac{1}{s_i}\right) s_i \, \mathrm{d}\mathcal{H}^{n-1} \leq n W_{n-i} \frac{F\left(\frac{|B_n|}{W_{n-i}} + \bar{\eta}_{i,i+1}\right) + F\left(\frac{|B_n|}{W_{n-i}} - \bar{\eta}_{i,i+1}\right)}{2},$$

where now, for any $0 \leq j, k \leq n$,

$$\bar{\eta}_{j,k} = \begin{cases} \dfrac{W_{n-j}W_{n-1} - W_{n-j-1}|B_n|}{W_{n-k+1}^2 \omega_a} & \text{if } K \neq rB_n \text{ for all } r > 0, \\ 0 & \text{if } K = rB_n. \end{cases}$$

Equality holds in both inequalities if $K = B_n$ (up to dilations).

Proof In order to get the first inequality we apply Proposition 1 to the probability space $\left(\mathbb{S}^{n-1}, \mathcal{H}^{n-1}/(n|B_n|)\right)$ and to the functions $\rho = s_i$ and $h = h_K$. Then, using the Minkowski integral formula (17), we get

$$\mathbb{E}\rho = \frac{W_{n-i}}{|B_n|}, \qquad \mathbb{E}h = \frac{W_{n-1}}{|B_n|}$$

and

$$\mathrm{Cov}(\rho, h) = \frac{W_{n-i-1}|B_n| - W_{n-i}W_{n-1}}{|B_n|^2}.$$

In addition,

$$\|h - \mathbb{E}h\|_\infty = \sup \left\{ \left| h_K(u) - \frac{W_{n-1}}{|B_n|} \right| : u \in \mathbb{S}^{n-1} \right\}$$

$$= \max \left\{ \bar{R} - \frac{W_{n-1}}{|B_n|}, \frac{W_{n-1}}{|B_n|} - \bar{r} \right\},$$

and since the functionals s_j, W_j are translation invariant, the smallest possible upper bound for $\int_{\mathrm{bd}\,K}(F \circ s_i)\,\mathrm{d}\mathscr{H}^{n-1}$ will be obtained for the translation of K such that the above maximum is minimal.

Moreover, using (4) we have

$$\frac{\mathrm{Cov}(\rho, h)}{\|h - \mathbb{E}h\|_\infty} = \frac{W_{n-i-1}|B_n| - W_{n-i}W_{n-1}}{|B_n|^2 \min_{x \in K} \max \left\{ \bar{R}(K-x) - \frac{W_{n-1}}{|B_n|}, \frac{W_{n-1}}{|B_n|} - \bar{r}(K-x) \right\}}$$

$$\geq \frac{W_{n-i-1}|B_n| - W_{n-i}W_{n-1}}{|B_n|^2 \min_{x \in K}\{\bar{R}(K-x) - \bar{r}(K-x)\}} = \frac{W_{n-i-1}|B_n| - W_{n-i}W_{n-1}}{|B_n|^2 \omega_a} = -\bar{\eta}_{i,1}.$$

Altogether and the concavity of F show the first inequality.

Second inequality is obtained analogously, but now as a consequence of Proposition 2 for $\rho = 1/s_i$, $h = h_K$ and $g = s_i/(nW_{n-i})$; we notice that, by (17), $\int_{\mathbb{S}^{n-1}} g\,\mathrm{d}\mathscr{H}^{n-1} = 1$. The equality case is trivial. □

If we replace $F(x)$ by the concave function x^α, $0 \leq \alpha \leq 1$, we get the corresponding result to Theorem 4 for the s_i's.

Theorem 6 *Let $K \in \mathscr{K}_0^n$ be of class C_+^2. Then, for all $i = 0, \ldots, n-1$, the following inequalities hold:*

- *If $0 \leq \alpha \leq 1/2$,*

$$\int_{\mathbb{S}^{n-1}} s_i^\alpha\,\mathrm{d}\mathscr{H}^{n-1} \leq \frac{n}{2}|B_n|^{1-\alpha}\left(W_{n-i} + |B_n|\bar{\eta}_{i,1}\right)^\alpha + \left(W_{n-i} - |B_n|\bar{\eta}_{i,1}\right)^\alpha.$$

- *If $1/2 \leq \alpha \leq 1$,*

$$\int_{\mathbb{S}^{n-1}} s_i^\alpha\,\mathrm{d}\mathscr{H}^{n-1} \leq \frac{n}{2}W_{n-i}^\alpha\left[\left(|B_n| + W_{n-i}\,\bar{\eta}_{i,i+1}\right)^{1-\alpha} + \left(|B_n| - W_{n-i}\,\eta_{i,i+1}\right)^{1-\alpha}\right].$$

Equality holds in both inequalities if $K = B_n$ (up to dilations).

5 Another Consequence: The Radial Function and the Dual Quermaßintegrals

In this section we will apply Proposition 1 in a difference setting: instead of working with convex bodies we will consider the so-called starshaped sets. A non-empty set $S \subset \mathbb{R}^n$ is called *starshaped* (with respect to the origin) if the line segment $[0, x] \subseteq S$ for all $x \in S$. For a compact starshaped set K, the radial function is defined as

$$\rho_K(u) = \max \{\lambda \geq 0 : \lambda u \in K\}, \quad u \in \mathbb{R}^n \setminus \{0\}.$$

Clearly, $\rho_K(u)u \in \text{bd } K$. We will denote by $\mathscr{S}_0^n$ the family of all compact starshaped sets in $\mathbb{R}^n$ having the origin as an interior point.

Closely related to the radial function are dual quermaßintegrals (and dual mixed volumes), which were introduced by Lutwak in [9]; they were the starting point for the development of the nowadays known as dual Brunn-Minkowski theory (see e.g. [12, Section 9.3]). For $K \in \mathscr{S}_0^n$ and $i = 0, \ldots, n$, the *dual quermaßintegral of order* $n - i$, $\widetilde{W}_{n-i}(K)$, is defined by

$$\widetilde{W}_{n-i}(K) = \frac{1}{n} \int_{\mathbb{S}^{n-1}} \rho_K^i \, d\mathscr{H}^{n-1}. \tag{18}$$

The functional $\widetilde{W}_{n-i}$ is non-negative, monotonous and homogeneous of degree i (see e.g. [7, Section A.7]), although it is not translation invariant. In particular, the use of spherical coordinates immediately yields $\widetilde{W}_0(K) = |K|$, whereas $\widetilde{W}_n(K) = |B_n|$ and $2\widetilde{W}_{n-1}(K)/|B_n|$ is the average length of chords of K through the origin. Moreover, if $K \in \mathscr{K}_0^n$ then $\widetilde{W}_i(K) \leq W_i(K)$ for all $i = 0, \ldots, n$ (see [9]).

For $K \in \mathscr{S}_0^n$, its in- and outer radii, $\tilde{r}(K)$, $\tilde{R}(K)$, are defined analogously to the convex case, and from the already mentioned monotonicity of the dual quermaßintegrals we get (cf. (4))

$$\tilde{r}(K)^{n-j} \widetilde{W}_{n-j+k}(K) \leq \widetilde{W}_k(K) \leq \tilde{R}(K)^{n-j} \widetilde{W}_{n-j+k}(K). \tag{19}$$

We observe that definition (18) can be extended to any real number, and thus, in contrast to the case of the classical quermaßintegrals, dual quermaßintegrals can be defined for any $i \in \mathbb{R}$. Now, the *dual Aleksandrov-Fenchel inequalities* (see e.g. [9, Theorem 2]) read (cf. (6))

$$\widetilde{W}_j(K)^{k-i} \leq \widetilde{W}_i(K)^{k-j} \widetilde{W}_k(K)^{j-i}, \quad \text{for } i \leq j \leq k. \tag{20}$$

In [1], a slightly stronger version of the following result was obtained. Again, for the sake of brevity, we will write $\widetilde{W}_i = \widetilde{W}_i(K)$.

Theorem 7 *Let $K \in \mathcal{S}_0^n$. For any convex function $F : I \longrightarrow \mathbb{R}, I \subseteq \mathbb{R}$ where all the quantities are defined, and all $i = 0, \ldots, n$,*

$$\int_{\mathbb{S}^{n-1}} \left(F \circ \rho_K^i \right) \mathrm{d}\mathcal{H}^{n-1} \geq n|B_n| \frac{F \left(\frac{\widetilde{W}_{n-i}}{|B_n|} + \widetilde{\eta}_{i,0} \right) + F \left(\frac{\widetilde{W}_{n-i}}{|B_n|} - \widetilde{\eta}_{i,0} \right)}{2}, \qquad (21)$$

$$\int_{\mathbb{S}^{n-1}} \left(F \circ \frac{1}{\rho_K^i} \right) \rho_K^i \, \mathrm{d}\mathcal{H}^{n-1} \geq n\widetilde{W}_{n-i} \frac{F \left(\frac{|B_n|}{\widetilde{W}_{n-i}} + \widetilde{\eta}_{i,i} \right) + F \left(\frac{|B_n|}{\widetilde{W}_{n-i}} - \widetilde{\eta}_{i,i} \right)}{2},$$

where now, for any $0 \leq j, k \leq n$,

$$\widetilde{\eta}_{j,k} = \frac{|K||B_n| - \widetilde{W}_{n-j}\widetilde{W}_j}{\widetilde{W}_{n-k}^2 \left(\bar{R}^{n-j} - \bar{r}^{n-j} \right)} \quad \text{if } K \neq rB_n, r > 0,$$

$$\widetilde{\eta}_{j,k} = 0 \qquad \text{if } K = rB_n \text{ for some } r > 0.$$

Equality holds in both inequalities if $K = B_n$ (up to dilations).

We observe that the relation $|K||B_2^n| \geq \widetilde{W}_{n-j}\widetilde{W}_j$, a consequence of the dual Aleksandrov-Fenchel inequality (20), ensures that $\widetilde{\eta}_{j,k} \geq 0$.

Following the same argument as in the proof of the above theorem, we obtain the corresponding result for the case of a concave function.

Theorem 8 *Let $K \in \mathcal{S}_0^n$. For any concave function $F : I \longrightarrow \mathbb{R}, I \subseteq \mathbb{R}$ where all the quantities are defined, and all $i = 0, \ldots, n$,*

$$\int_{\mathbb{S}^{n-1}} \left(F \circ \rho_K^i \right) \mathrm{d}\mathcal{H}^{n-1} \leq n|B_n| \frac{F \left(\frac{\widetilde{W}_{n-i}}{|B_n|} + \widetilde{\eta}_{i,0} \right) + F \left(\frac{\widetilde{W}_{n-i}}{|B_n|} - \widetilde{\eta}_{i,0} \right)}{2}, \qquad (22)$$

$$\int_{\mathbb{S}^{n-1}} \left(F \circ \frac{1}{\rho_K^i} \right) \rho_K^i \, \mathrm{d}\mathcal{H}^{n-1} \leq n\widetilde{W}_{n-i} \frac{F \left(\frac{|B_n|}{\widetilde{W}_{n-i}} + \widetilde{\eta}_{i,i} \right) + F \left(\frac{|B_n|}{\widetilde{W}_{n-i}} - \widetilde{\eta}_{i,i} \right)}{2}.$$

Equality holds in both inequalities if $K = B_n$ (up to dilations).

Proof In order to prove (22) we apply Proposition 1 to the probability space $\left(\mathbb{S}^{n-1}, \mathcal{H}^{n-1}/(n|B_n|) \right)$ and the functions $\rho = \rho_K^i$ and $h = \rho_K^{n-i}$. Then, using (18), $\mathbb{E}\rho = \widetilde{W}_{n-i}/|B_n|$, $\mathbb{E}h = \widetilde{W}_i/|B_n|$ and

$$\mathrm{Cov}(\rho, h) = \frac{|K||B_n| - \widetilde{W}_{n-i}\widetilde{W}_i}{|B_n|^2}.$$

Moreover, since $\rho_K(u)u \in \mathrm{bd}\,K$, the relations (19) yield

$$\|h - \mathbb{E}h\|_\infty = \sup\left\{\left|\rho_K(u)^{n-i} - \frac{\widetilde{W}_i}{|B_n|}\right| : u \in \mathbb{S}^{n-1}\right\}$$

$$= \max\left\{\bar{R}^{n-i} - \frac{\widetilde{W}_i}{|B_n|}, \frac{\widetilde{W}_i}{|B_n|} - \bar{r}^{n-i}\right\} \geq \bar{R}^{n-i} - \bar{r}^{n-i}.$$

Altogether and the concavity of F show the first inequality.

Second inequality is obtained analogously, but now as a consequence of Proposition 2 for $\rho = 1/\rho_K^i$, $h = \rho_K^{n-i}$ and $g = \rho_K^i/(n\widetilde{W}_{n-i})$. The equality case is trivial. $\quad\square$

We observe that since (18) can be defined for any $i \in \mathbb{R}$, taking $F(x) = x^\alpha$ or $F(x) = 1/x^\alpha$ for suitable powers $\alpha \geq 0$, new inequalities relating the dual quermaßintegrals with the in- and outer radii can be obtained. Indeed, even Theorems 7 and 8 hold true for all $i \in \mathbb{R}$, just properly defining the values $\widetilde{\eta}_{j,k}$.

For instance, taking $F(x) = x^2$ in (21), then

$$\int_{\mathbb{S}^{n-1}} \left(F \circ \rho_K^i\right) d\mathscr{H}^{n-1} = \int_{\mathbb{S}^{n-1}} \rho_K^{2i} d\mathscr{H}^{n-1} = n\widetilde{W}_{n-2i}$$

for any $i = 0, \ldots, n$, and hence we get

$$|B_n|^2 \left(|B_n|\widetilde{W}_{n-2i} - \widetilde{W}_{n-i}^2\right)\left(\bar{R}^{n-i} - \bar{r}^{n-i}\right)^2 \geq \left(|K||B_n| - \widetilde{W}_{n-i}\widetilde{W}_i\right)^2,$$

with equality for the ball.

If we consider now the concave function $F(x) = \sqrt{x}$ and apply (22), we obtain

$$2\widetilde{W}_{n-i/2}^2 \leq |B_n|\widetilde{W}_{n-i} + \sqrt{|B_n|^2\widetilde{W}_{n-i}^2 - \frac{\left(|K||B_n| - \widetilde{W}_{n-i}\widetilde{W}_i\right)^2}{\left(\bar{R}^{n-i} - \bar{r}^{n-i}\right)}};$$

here we are assuming that $K \neq rB_n$, otherwise we get a trivial identity.

Acknowledgements First author is partially supported by MINECO project MTM2016-77710-P. Second author is supported by MINECO/FEDER project MTM2015-65430-P and "Programa de Ayudas a Grupos de Excelencia de la Región de Murcia", Fundación Séneca, 19901/GERM/15.

References

1. D. Alonso-Gutiérrez, M.A. Hernández Cifre, A.R. Martínez Fernández, I.: Bounding the integral of powered i-th mean curvatures. Rev. Mat. Iberoamericana **33**(4), 1197–1218 (2017)
2. I. Bárány, On the minimal ring containing the boundary of a convex body. Acta Sci. Math. **52**, 93–100 (1988)

3. B.-Y. Chen, On the total curvature of immersed manifolds. I. An inequality of Fenchel-Borsuk-Willmore. Am. J. Math. **93**, 148–162 (1971)
4. B.-Y. Chen, *Geometry of Submanifolds and Its Applications* (Science University of Tokyo, Tokyo, 1981)
5. M.P. do Carmo, *Differential Geometry of Curves and Surfaces* (Prentice-Hall, Englewood Cliffs, NJ, 1976)
6. M.E. Gage, An isoperimetric inequality with applications to curve shortening. Duke Math. J. **50**(4), 1225–1229 (1983)
7. R.J. Gardner, *Geometric Tomography*. Encyclopedia of Mathematics and its Applications, vol. 58, 2nd edn. (Cambridge University Press, Cambridge, 2006)
8. M. Green, S. Osher, Steiner polynomials, Wulff flows, and some new isoperimetric inequalities for convex plane curves. Asian J. Math. **3**(3), 659–676 (1999)
9. E. Lutwak, Dual mixed volumes. Pac. J. Math. **58**(2), 531–538 (1975)
10. M. Ritoré, C. Sinestrari, *Mean Curvature Flow and Isoperimetric Inequalities* (Birkhäuser, Basel, 2010)
11. A. Ros, Compact hypersurfaces with constant higher order mean curvatures. Rev. Mat. Iberoamericana **3**(3–4), 447–453 (1987)
12. R. Schneider, *Convex Bodies: The Brunn-Minkowski Theory*, 2nd expanded edition (Cambridge University Press, Cambridge, 2014)
13. J. Steiner, Über parallele Flächen. Monatsber. Preuss. Akad. Wiss. 114–118 (1840) [Ges. Werke, Vol II (Reimer, Berlin, 1882), pp. 245–308].
14. T.J. Willmore, Mean curvature of immersed surfaces. An. Şti. Univ. "All. I. Cuza" Iaşi Secţ. I a Mat. **14**, 99–103 (1968)

Crofton Formulae for Tensorial Curvature Measures: The General Case

Daniel Hug and Jan A. Weis

Abstract The tensorial curvature measures are tensor-valued generalizations of the curvature measures of convex bodies. On convex polytopes, there exist further generalizations some of which also have continuous extensions to arbitrary convex bodies. In a previous work, we obtained kinematic formulae for all (generalized) tensorial curvature measures. As a consequence of these results, we now derive a complete system of Crofton formulae for such (generalized) tensorial curvature measures. These formulae express the integral mean of the (generalized) tensorial curvature measures of the intersection of a given convex body (resp. polytope, or finite unions thereof) with a uniform affine k-flat in terms of linear combinations of (generalized) tensorial curvature measures of the given convex body (resp. polytope, or finite unions thereof). The considered generalized tensorial curvature measures generalize those studied formerly in the context of Crofton-type formulae, and the coefficients involved in these results are substantially less technical and structurally more transparent than in previous works. Finally, we prove that essentially all generalized tensorial curvature measures on convex polytopes are linearly independent. In particular, this implies that the Crofton formulae which we prove in this contribution cannot be simplified further.

1 Introduction

The *classical Crofton formula* is a major result in integral geometry. It expresses the integral mean of the intrinsic volume of a convex body intersected with a uniform affine subspace of the underlying Euclidean space in terms of another intrinsic volume of this convex body. More precisely, for a *convex body* $K \in \mathcal{K}^n$ (a nonempty, compact, convex set) in the n-dimensional Euclidean space $\mathbb{R}^n$, $n \in \mathbb{N}$,

D. Hug (✉) · J. A. Weis
Department of Mathematics, Karlsruhe Institute of Technology, Karlsruhe, Germany
e-mail: daniel.hug@kit.edu; jan.weis@kit.edu

© Springer International Publishing AG 2018
G. Bianchi et al. (eds.), *Analytic Aspects of Convexity*, Springer INdAM Series 25,
https://doi.org/10.1007/978-3-319-71834-7_3

the classical Crofton formula (see [26, (4.59)]) states that

$$\int_{A(n,k)} V_j(K \cap E)\, \mu_k(dE) = \alpha_{njk} V_{n-k+j}(K),\tag{1}$$

for $k \in \{0, \ldots, n\}$ and $j \in \{0, \ldots, k\}$, where $A(n, k)$ is the affine Grassmannian of k-flats in $\mathbb{R}^n$, on which μ_k denotes the motion invariant Haar measure, normalized as in [27, p. 588], and

$$\alpha_{njk} = \frac{\Gamma(\frac{n-k+j+1}{2})\Gamma(\frac{k+1}{2})}{\Gamma(\frac{n+1}{2})\Gamma(\frac{j+1}{2})}$$

is expressed in terms of specific values of the Gamma function $\Gamma(\cdot)$ (see [26, Theorem 4.4.2]).

The functionals $V_i : \mathcal{K}^n \to \mathbb{R}$, for $i \in \{0, \ldots, n\}$, appearing in (1), are the *intrinsic volumes*, which occur as the coefficients of the monomials in the *Steiner formula*

$$\mathcal{H}^n(K + \epsilon B^n) = \sum_{j=0}^{n} \kappa_{n-j} V_j(K) \epsilon^{n-j},\tag{2}$$

which holds for all convex bodies $K \in \mathcal{K}^n$ and $\epsilon \geq 0$. Here, $\mathcal{H}^n$ is the n-dimensional Hausdorff measure (Lebesgue measure, volume), $+$ denotes the Minkowski addition in $\mathbb{R}^n$, and κ_n is the volume of the Euclidean unit ball B^n in $\mathbb{R}^n$. Properties of the intrinsic volume V_i such as continuity, isometry invariance, homogeneity, and additivity (valuation property) are derived from corresponding properties of the volume functional. A key result for the intrinsic volumes is *Hadwiger's characterization theorem* (see [7, 2. Satz]), which states that $V_0, \ldots, V_n$ form a basis of the vector space of continuous and isometry invariant real-valued valuations on $\mathcal{K}^n$. One of its numerous applications is a concise proof of (1).

A natural way to extend the classical Crofton formula is to apply the integration over the affine Grassmannian $A(n, k)$ to functionals which generalize the intrinsic volumes. One of these generalizations concerns tensor-valued valuations on $\mathcal{K}^n$. Their systematic investigation started with a characterization theorem, similar to the aforementioned result due to Hadwiger. Integral geometric formulae, including a Crofton formula, for *quermassvectors* (vector-valued generalizations of the intrinsic volumes) have already been found by Hadwiger & Schneider and Schneider, in 1971/1972 (see [8, 21, 22]). More recently in 1997, McMullen generalized these vector-valued valuations even further, and introduced tensor-valued generalizations of the intrinsic volumes (see [18]). Only 2 years later Alesker generalized Hadwiger's characterization theorem (see [1, Theorem 2.2]) by showing that the vector space of continuous and isometry covariant tensor-valued valuations on $\mathcal{K}^n$ is spanned by the tensor-valued versions of the intrinsic volumes, the *Minkowski tensors*, multiplied with suitable powers of the metric tensor in $\mathbb{R}^n$. However, these valuations are not linearly independent, as shown by McMullen (see [18]) and further investigated by Hug et al. (see [16]). This is one reason why an approach to explicit integral geometric formulae via characterization theorems does not seem to be technically feasible. Nevertheless, great progress in the integral geometry of tensor-valued valuations has been made by different methods. In 2008, Hug et al. proved a set of Crofton formulae for the Minkowski tensors (see [15, Theorem 2.1–2.6]). A totally different algebraic approach has been developed by Bernig and Hug to obtain various integral geometric formulae for the translation invariant Minkowski tensors (see [3]).

On the other hand, localizations of the intrinsic volumes yield other types of generalizations. The *support measures* are weakly continuous, locally defined and motion equivariant valuations on convex bodies with values in the space of finite measures on Borel subsets of $\mathbb{R}^n \times \mathbb{S}^{n-1}$, where $\mathbb{S}^{n-1}$ denotes the Euclidean unit sphere in $\mathbb{R}^n$. They are determined by a local version of (2) and form a crucial example of localizations of the intrinsic volumes, which are simply the total support measures. Furthermore, their marginal measures on Borel subsets of $\mathbb{R}^n$ are called *curvature measures*, and the ones on Borel subsets of $\mathbb{S}^{n-1}$ are called *area measures*. For the area measures and the curvature measures, Schneider found characterization theorems (see [23, 24]) similar to the one due to Hadwiger in the global case. It took some time until in 1995, Glasauer proved a characterization theorem for the support measures, even without the need of requesting the valuation property (see [5, Satz 4.2.1]). As to integral geometry, in 1959 Federer [4] proved Crofton formulae for curvature measures, even in the more general setting of sets with positive reach. Certain Crofton formulae for support measures were proved by Glasauer in 1997 (see [6, Theorem 3.2]). However, his results require a special set operation on support elements of the involved convex bodies and affine subspaces.

Interestingly, the combination of Minkowski tensors and localization leads to a better understanding of integral geometric formulae. In recent years, Schneider introduced *local tensor valuations* (see [25]), which were then further studied by Hug and Schneider (see [9–11]). They introduced particular tensor-valued support measures, the *local Minkowski tensors* on convex bodies (and generalizations on polytopes), which (as their name suggests) can also be seen as localizations of the Minkowski tensors. They proved several different characterization results for these (generalized) local Minkowski tensors in the just mentioned works. This led us to consider their marginal measures on Borel subsets of $\mathbb{R}^n$, the *tensorial curvature measures* and their generalizations on convex polytopes. Preceding this work, the present authors derived a set of Crofton formulae for a different version of these tensorial curvature measures, defined with respect to the (random) intersecting affine subspace, and as a consequence of these results also obtained Crofton formulae for some of the (original) tensorial curvature measures (see [13]). As a far reaching generalization of previous results, a complete set of kinematic formulae for the (generalized) tensorial curvature measures has been proved in [12].

In statistical physics, the intrinsic volumes (in the physical context better known as *Minkowski functionals*) are an important tool for the characterization of geometric properties of spatial patterns (see for example the survey [19]). However, due to their translation and rotation invariance, they are not useful when it comes to the quantification of orientation or anisotropy of spatial structures. For the determination of these kinds of geometric features, rotation covariant valuations, such as the Minkowski tensors and the (generalized) tensorial curvature measures considered here, are much more suitable, and therefore they have been heavily used recently (see for example [28]). For further applications we refer to the introduction of our preceding work [12].

The aim of the present work is to prove a complete set of Crofton formulae for the (generalized) tensorial curvature measures. This complements the particular results for (extrinsic) tensorial curvature measures and Minkowski tensors obtained in [13] and [29]. The current approach is basically an application of the kinematic formulae for (generalized) tensorial curvature measures derived in [12]. The connection between local kinematic and local Crofton formulae is well known for the scalar curvature measures. In that case it is used to determine the coefficients in the kinematic formulae. The basic strategy there is as follows. First, the kinematic formulae are proved, but the involved coefficients remain undetermined, since the required direct calculation seemed to be infeasible. Then Crofton formulae are derived which involve the same constants. In the latter, the determination of the coefficients turns out to be an easy task, which is accomplished by evaluating the result for balls of different radii. In the tensorial framework, this approach breaks down, since the explicit calculation of integral mean values of (generalized) tensorial curvature measures for sufficiently many examples (template method) does not seem to be possible. Instead, the required coefficients were determined by a direct derivation of the kinematic formulae for (generalized) tensorial curvature measures in [12], and from this we now can derive explicit Crofton formulae, otherwise following the reasoning described above.

Since the tensorial curvature measures are local versions of the Minkowski tensors, it is rather straightforward to derive Crofton formulae, similar to the ones proved in the present contribution, and apparently also kinematic formulae, similar to the ones obtained in the preceding work [12], for Minkowski tensors as well. This is the subject of the subsequent work [14]. There we extend some of the integral formulae for translation invariant Minkowski tensors obtained by Bernig and Hug in [3] and significantly simplify the coefficients of the Crofton formulae proven in [15]. We further refer to [12] for a thorough discussion of related work.

The present contribution is structured as follows. In Sect. 2, we fix our notation and collect various auxiliary results which will be needed. Section 3 contains the main results. First, we state the Crofton formulae for the generalized tensorial curvature measures on the space $\mathscr{P}^n$ of convex polytopes in $\mathbb{R}^n$. Then we provide the formulae for all the (generalized) tensorial curvature measures for which a continuous extension to $\mathscr{K}^n$ exists. Finally, we highlight some special cases. In Sect. 4, we first recall the kinematic formulae for generalized tensorial curvature measures from [12], in order to apply these in the proofs of the main results and the corollaries. In the final Sect. 5 we show that the generalized tensorial curvature measures on convex polytopes are essentially all linearly independent.

2 Preliminaries

We work in the n-dimensional Euclidean space $\mathbb{R}^n$, equipped with its usual topology generated by the standard scalar product $\langle \cdot, \cdot \rangle$ and the corresponding Euclidean norm $\| \cdot \|$. For a topological space X, we denote the Borel σ-algebra on X by $\mathscr{B}(X)$.

We denote the proper (orientation preserving) rotation group on $\mathbb{R}^n$ by $SO(n)$, and we write ν for the Haar probability measure on $SO(n)$. By $G(n,k)$ (resp. $A(n,k)$), for $k \in \{0, \ldots, n\}$, we denote the Grassmannian of k-dimensional linear (resp. affine) subspaces of $\mathbb{R}^n$. These sets carry natural topologies, see [27, Chap. 13.2].

We write μ_k for the rotation invariant Haar measure on $A(n,k)$, normalized as in [27, (13.2)], that is, for a fixed (but arbitrary) linear subspace $E_k \in G(n,k)$,

$$\mu_k(\cdot) = \int_{SO(n)} \int_{E_k^{\perp}} \mathbf{1}\{\rho(E_k + t) \in \cdot\}\, \mathscr{H}^{n-k}(\mathrm{d}t)\, \nu(\mathrm{d}\rho), \tag{3}$$

where $\mathscr{H}^j$ denotes the j-dimensional Hausdorff measure for $j \in \{0, \ldots, n\}$. The directional space of an affine k-flat $E \in A(n,k)$ is denoted by $E^0 \in G(n,k)$ and its orthogonal complement by $E^{\perp} \in G(n, n-k)$. The orthogonal projection of a vector $x \in \mathbb{R}^n$ to a linear subspace L of $\mathbb{R}^n$ is denoted by $p_L(x)$.

The vector space of symmetric tensors of rank $p \in \mathbb{N}_0$ over $\mathbb{R}^n$ is denoted by $\mathbb{T}^p$, and the corresponding algebra of symmetric tensors over $\mathbb{R}^n$ by $\mathbb{T}$. The symmetric tensor product of two tensors $T, U \in \mathbb{T}$ is denoted by TU, and for $q \in \mathbb{N}_0$ and a tensor $T \in \mathbb{T}$ we write T^q for the q-fold tensor product; see also the introductory chapters of [17] for further details and references. Identifying $\mathbb{R}^n$ with its dual space via its scalar product, we interpret a symmetric tensor of rank p as a symmetric p-linear map from $(\mathbb{R}^n)^p$ to $\mathbb{R}$. One special tensor is the *metric tensor* $Q \in \mathbb{T}^2$, defined by $Q(x,y) := \langle x, y \rangle$ for $x, y \in \mathbb{R}^n$. For an affine k-flat $E \in A(n,k), k \in \{0, \ldots, n\}$, the metric tensor $Q(E)$ associated with E is defined by $Q(E)(x,y) := \langle p_{E^0}(x), p_{E^0}(y) \rangle$ for $x, y \in \mathbb{R}^n$.

The definition of the (generalized) tensorial curvature measures, which we recall in the following, is partly motivated by their relation to the support measures. Therefore, we first recall the latter. For a convex body $K \in \mathscr{K}^n$ and $x \in \mathbb{R}^n$, we denote the metric projection of x onto K by $p(K,x)$, and for $x \in \mathbb{R}^n \setminus K$ we define $u(K,x) := (x - p(K,x))/\|x - p(K,x)\|$. For $\epsilon > 0$ and a Borel set $\eta \subset \Sigma^n := \mathbb{R}^n \times \mathbb{S}^{n-1}$,

$$M_\epsilon(K, \eta) := \{x \in (K + \epsilon B^n) \setminus K : (p(K,x), u(K,x)) \in \eta\}$$

is a local parallel set of K which satisfies the *local Steiner formula*

$$\mathscr{H}^n(M_\epsilon(K, \eta)) = \sum_{j=0}^{n-1} \kappa_{n-j} \Lambda_j(K, \eta) \epsilon^{n-j}, \qquad \epsilon \geq 0. \tag{4}$$

This relation determines the *support measures* $\Lambda_0(K, \cdot), \ldots, \Lambda_{n-1}(K, \cdot)$ of K, which are finite, nonnegative Borel measures on $\mathscr{B}(\Sigma^n)$. Obviously, a comparison of (4) and the Steiner formula (2) yields $V_j(K) = \Lambda_j(K, \Sigma^n)$. Further information on these measures and functionals can be found in [26, Chap. 4.2].

Let $\mathscr{P}^n \subset \mathscr{K}^n$ denote the space of convex polytopes in $\mathbb{R}^n$. For a polytope $P \in \mathscr{P}^n$ and $j \in \{0, \ldots, n\}$, we denote the set of j-dimensional faces of P by $\mathscr{F}_j(P)$ and the normal cone of P at a face $F \in \mathscr{F}_j(P)$ by $N(P, F)$. Then, the jth support measure $\Lambda_j(P, \cdot)$ of P is explicitly given by

$$\Lambda_j(P, \eta) = \frac{1}{\omega_{n-j}} \sum_{F \in \mathscr{F}_j(P)} \int_F \int_{N(P,F) \cap \mathbb{S}^{n-1}} \mathbf{1}_\eta(x, u) \, \mathscr{H}^{n-j-1}(\mathrm{d}u) \, \mathscr{H}^j(\mathrm{d}x)$$

for $\eta \in \mathscr{B}(\Sigma^n)$ and $j \in \{0, \ldots, n-1\}$, where ω_n is the $(n-1)$-dimensional volume (Hausdorff measure) of $\mathbb{S}^{n-1}$.

For a polytope $P \in \mathscr{P}^n$, we define the *generalized tensorial curvature measure*

$$\phi_j^{r,s,l}(P, \cdot), \qquad j \in \{0, \ldots, n-1\}, \, r, s, l \in \mathbb{N}_0,$$

as the Borel measure on $\mathscr{B}(\mathbb{R}^n)$ which is given by

$$\phi_j^{r,s,l}(P, \beta) := c_{n,j}^{r,s,l} \frac{1}{\omega_{n-j}} \sum_{F \in \mathscr{F}_j(P)} Q(F)^l \int_{F \cap \beta} x^r \, \mathscr{H}^j(\mathrm{d}x) \int_{N(P,F) \cap \mathbb{S}^{n-1}} u^s \, \mathscr{H}^{n-j-1}(\mathrm{d}u),$$

for $\beta \in \mathscr{B}(\mathbb{R}^n)$, where

$$c_{n,j}^{r,s,l} := \frac{1}{r! s!} \frac{\omega_{n-j}}{\omega_{n-j+s}} \frac{\omega_{j+2l}}{\omega_j} \text{ if } j \neq 0, \; c_{n,0}^{r,s,0} := \frac{1}{r! s!} \frac{\omega_n}{\omega_{n+s}}, \text{ and } c_{n,0}^{r,s,l} := 1 \text{ for } l \geq 1.$$

Note that if $j = 0$ and $l \geq 1$, then we have $\phi_0^{r,s,l} \equiv 0$. In all other cases the factor $1/\omega_{n-j}$ in the definition of $\phi_j^{r,s,l}(P, \beta)$ and the factor ω_{n-j} involved in the constant $c_{n,j}^{r,s,l}$ cancel.

For a general convex body $K \in \mathscr{K}^n$, we define the *tensorial curvature measure*

$$\phi_n^{r,0,l}(K, \cdot), \qquad r, l \in \mathbb{N}_0,$$

as the Borel measure on $\mathscr{B}(\mathbb{R}^n)$ which is given by

$$\phi_n^{r,0,l}(K, \beta) := c_{n,n}^{r,0,l} Q^l \int_{K \cap \beta} x^r \, \mathscr{H}^n(\mathrm{d}x),$$

for $\beta \in \mathscr{B}(\mathbb{R}^n)$, where $c_{n,n}^{r,0,l} := \frac{1}{r!} \frac{\omega_{n+2l}}{\omega_n}$. For the sake of convenience, we extend these definitions by $\phi_j^{r,s,0} := 0$ for $j \notin \{0, \ldots, n\}$ or $r \notin \mathbb{N}_0$ or $s \notin \mathbb{N}_0$ or $j = n$ and $s \neq 0$. Finally, we observe that for $P \in \mathscr{P}^n$, $r = s = l = 0$, and $j = 0, \ldots, n-1$, the scalar-valued measures $\phi_j^{0,0,0}(P, \cdot)$ are just the curvature measures $\phi_j(P, \cdot)$, that is, the marginal measures on $\mathbb{R}^n$ of the support measures $\Lambda_j(P, \cdot)$, which therefore can be extended from polytopes to general convex bodies, and $\phi_n^{0,0,0}(K, \cdot)$ is the restriction of the n-dimensional Hausdorff measure to $K \in \mathcal{K}^n$.

We emphasize that in the present work, $\phi_j^{r,s,l}(P, \cdot)$ and $\phi_n^{r,0,l}(K, \cdot)$ are Borel measures on $\mathbb{R}^n$ and not on $\mathbb{R}^n \times \mathbb{S}^{n-1}$, as in [9], and also the normalization is slightly adjusted as compared to [9] (where the normalization was not a relevant issue). However, we stick to the definition and normalization of our preceding work [12], where the connection to the generalized local Minkowski tensors $\tilde{\phi}_j^{r,s,l}$ is described and where also the properties and available characterization results for these measures are discussed in more detail.

It has been shown in [9] that the generalized local Minkowski tensor $\tilde{\phi}_j^{r,s,l}$ has a continuous extension to $\mathcal{K}^n$ which preserves all other properties if and only if $l \in \{0, 1\}$; see [9, Theorem 2.3] for a stronger characterization result. Globalizing any such continuous extension in the $\mathbb{S}^{n-1}$-coordinate, we obtain a continuous extension for the generalized tensorial curvature measures. These can be represented with suitable differential forms which are defined on the sphere bundle of $\mathbb{R}^n$ and integrated (that is, evaluated) on the normal cycle, which is the reason why they are called *smooth* (for more details, see for example [9, 17, 20] and the literature cited there). For $l = 0$, this extension can be easily expressed via the support measures. We call the measures thus obtained the *tensorial curvature measures*. For a convex body $K \in \mathcal{K}^n$, a Borel set $\beta \in \mathcal{B}(\mathbb{R}^n)$, and $r, s \in \mathbb{N}_0$, they are given by

$$\phi_j^{r,s,0}(K, \beta) := c_{n,j}^{r,s,0} \int_{\beta \times \mathbb{S}^{n-1}} x^r u^s \, \Lambda_j(K, \mathrm{d}(x, u)), \tag{5}$$

for $j \in \{0, \ldots, n-1\}$, whereas $\phi_n^{r,0,l}(K, \beta)$ has already been defined for all $K \in \mathcal{K}^n$.

The valuations $Q^m \phi_j^{r,s,l}$ on $\mathcal{P}^n$ are linearly independent, where $m, j, r, s, l \in \mathbb{N}_0$, $j \in \{0, \ldots, n\}$ with $l = 0$, if $j \in \{0, n-1\}$, and $s = l = 0$, if $j = n$. In our preceding work [12], we pointed out that the corresponding proof is similar to the proof of [9, Theorem 3.1]. A detailed argument is now provided in Sect. 5.

The statements of our results involve the classical *Gamma function*. For all $z \in \mathbb{C} \setminus \{0, -1, \ldots\}$ (see [2, (2.7)]), it can be defined via the Gaussian product formula

$$\Gamma(z) := \lim_{a \to \infty} \frac{a^z a!}{z(z+1) \cdots (z+a)}.$$

This definition implies, for $c \in \mathbb{R} \setminus \mathbb{Z}$ and $m \in \mathbb{N}_0$, that

$$\frac{\Gamma(-c+m)}{\Gamma(-c)} = (-1)^m \frac{\Gamma(c+1)}{\Gamma(c-m+1)}. \tag{6}$$

The Gamma function has simple poles at the nonpositive integers. The right side of relation (6) provides a continuation of the left side at $c \in \mathbb{N}_0$, with the understanding that $\Gamma(c - m + 1)^{-1} = 0$ for $c < m$. This continuation will be applied several times (sometimes without mentioning it specifically) in the coefficients of the upcoming formulae and the corresponding proofs. Mostly it is used in "boundary cases" in which quotients are involved, such as in (6) with $c = 0$ and $m \in \mathbb{N}_0$, which have to be interpreted as $\Gamma(0 + m)/\Gamma(0) = \mathbf{1}\{m = 0\}$.

3   The Crofton Formulae

In this work, we establish a complete set of Crofton formulae for the generalized tensorial curvature measures of convex polytopes. That is, for $P \in \mathscr{P}^n$ and $\beta \in \mathscr{B}(\mathbb{R}^n)$, we explicitly express integrals of the form

$$\int_{A(n,k)} \phi_j^{r,s,l}(P \cap E, \beta \cap E)\, \mu_k(dE)$$

in terms of generalized tensorial curvature measures of P, evaluated at β. Furthermore, for $l = 0, 1$, the corresponding (generalized) tensorial curvature measures are defined on $\mathscr{K}^n \times \mathscr{B}(\mathbb{R}^n)$, and therefore we also consider the Crofton integrals

$$\int_{A(n,k)} \phi_j^{r,s,l}(K \cap E, \beta \cap E)\, \mu_k(dE)$$

for $K \in \mathscr{K}^n$, $\beta \in \mathscr{B}(\mathbb{R}^n)$, $l = 0, 1$.

All results which are stated in the following, extend by additivity to finite unions of polytopes or convex bodies.

3.1   Generalized Tensorial Curvature Measures on Polytopes

First, we separately state a formula for $j = k$.

Theorem 1 *Let* $P \in \mathscr{P}^n$, $\beta \in \mathscr{B}(\mathbb{R}^n)$, *and* $k, r, s, l \in \mathbb{N}_0$ *with* $k \leq n$. *Then,*

$$\int_{A(n,k)} \phi_k^{r,s,l}(P \cap E, \beta \cap E)\, \mu_k(dE) = \mathbf{1}\{s \text{ even}\} \frac{1}{(2\pi)^s \left(\frac{s}{2}\right)!} \frac{\Gamma(\frac{n-k+s}{2})}{\Gamma(\frac{n-k}{2})} \phi_n^{r,0,\frac{s}{2}+l}(P, \beta).$$

Theorem 1 generalizes Theorem 2.1 in [15]. In fact, setting $l = 0$ and $\beta = \mathbb{R}^n$ one obtains the known result for Minkowski tensors. If $l \in \{0, 1\}$, one can even formulate Theorem 1 for a convex body, as in both of these cases all appearing valuations are defined on $\mathscr{K}^n$. For $k = n$, the integral on the left-hand side of the formula in Theorem 1 is trivial. However, note that on the right-hand side the quotient of the Gamma functions has to be interpreted as $\mathbf{1}\{s = 0\}$, according to (6).

Next, we state the formulae for general $j < k$.

Theorem 2 *Let* $P \in \mathscr{P}^n$, $\beta \in \mathscr{B}(\mathbb{R}^n)$, *and* $j, k, r, s, l \in \mathbb{N}_0$ *with* $j < k \leq n$, *and with* $l = 0$ *if* $j = 0$. *Then,*

$$\int_{A(n,k)} \phi_j^{r,s,l}(P \cap E, \beta \cap E)\, \mu_k(dE) = \sum_{m=0}^{\lfloor \frac{s}{2} \rfloor} \sum_{i=0}^{m} d_{n,j,k}^{s,l,i,m}\, Q^{m-i} \phi_{n-k+j}^{r,s-2m,l+i}(P, \beta),$$

where

$$d_{n,j,k}^{s,l,i,m} := \frac{(-1)^i}{(4\pi)^m m!}\, \frac{\binom{m}{i}}{\pi^i}\, \frac{(i+l-2)!}{(l-2)!}\, \frac{\Gamma(\frac{n-k+j+1}{2})\Gamma(\frac{k+1}{2})}{\Gamma(\frac{n+1}{2})\Gamma(\frac{j+1}{2})}$$

$$\times\, \frac{\Gamma(\frac{n-k+j}{2}+1)}{\Gamma(\frac{n-k+j+s}{2}+1)}\, \frac{\Gamma(\frac{j+s}{2}-m+1)}{\Gamma(\frac{j}{2}+1)}\, \frac{\Gamma(\frac{n-k}{2}+m)}{\Gamma(\frac{n-k}{2})}.$$

For $k = n$ the coefficient in Theorem 2 has to be interpreted as

$$d_{n,j,n}^{s,l,i,m} = \mathbf{1}\{i = m = 0\},$$

according to (6), so that the result is a tautology in this case.

Several remarkable facts concerning the coefficients $d_{n,j,k}^{s,l,i,m}$ should be recalled from [12]. First, the ratio $(i+l-2)!/(l-2)!$ has to be interpreted in terms of Gamma functions and relation (6) if $l \in \{0,1\}$. The corresponding special cases will be considered separately in the following two theorems and the subsequent corollaries. Second, the coefficients are independent of the tensorial parameter r and, due to our normalization of the generalized tensorial curvature measures, depend only on l through the ratio $(i+l-2)!/(l-2)!$. Third, only tensors $\phi_{n-k+j}^{r,s-2m,p}(P, \beta)$ with $p \geq l$ show up on the right side of the kinematic formula. Using Legendre's duplication formula, we could shorten the given expressions for the coefficients $d_{n,j,k}^{s,l,i,m}$ even further. However, the present form has the advantage of exhibiting that the factors in the second line cancel each other if $s = 0$ (and hence also $m = i = 0$). Furthermore, in general the coefficients are signed in contrast to the classical kinematic formula. We shall see below that for $l \in \{0,1\}$ all coefficients are nonnegative.

3.2   (Generalized) Tensorial Curvature Measures on Convex Bodies

The generalized tensorial curvature measures $\phi_j^{r,s,l}$ can be continuously extended to all convex bodies if $l \in \{0,1\}$. In these two cases, Theorem 1 holds for general convex bodies as well. For this reason, we restrict our attention to the cases where $j < k$ in the following. The next theorems are stated without a proof, as they basically follow from Theorem 2 and approximation of the given convex body by polytopes (using the weak continuity of the corresponding generalized tensorial curvature measures and the usual arguments needed to take care of exceptional positions).

We start with the formula for $l = 1$.

Theorem 3 *Let $K \in \mathcal{K}^n$, $\beta \in \mathcal{B}(\mathbb{R}^n)$, and $j, k, r, s \in \mathbb{N}_0$ with $0 < j < k \leq n$. Then,*

$$\int_{A(n,k)} \phi_j^{r,s,1}(K \cap E, \beta \cap E) \, \mu_k(dE) = \sum_{m=0}^{\lfloor \frac{s}{2} \rfloor} d_{n,j,k}^{s,1,0,m} \, Q^m \phi_{n-k+j}^{r,s-2m,1}(K, \beta),$$

where

$$d_{n,j,k}^{s,1,0,m} = \frac{1}{(4\pi)^m m!} \frac{\Gamma(\frac{n-k+j+1}{2}) \Gamma(\frac{k+1}{2})}{\Gamma(\frac{n+1}{2}) \Gamma(\frac{j+1}{2})}$$

$$\times \frac{\Gamma(\frac{n-k+j}{2} + 1)}{\Gamma(\frac{n-k+j+s}{2} + 1)} \frac{\Gamma(\frac{j+s}{2} - m + 1)}{\Gamma(\frac{j}{2} + 1)} \frac{\Gamma(\frac{n-k}{2} + m)}{\Gamma(\frac{n-k}{2})}.$$

Next, we state the formula for $l = 0$.

Theorem 4 *Let $K \in \mathcal{K}^n$, $\beta \in \mathcal{B}(\mathbb{R}^n)$ and $j, k, r, s \in \mathbb{N}_0$ with $j < k \leq n$. Then,*

$$\int_{A(n,k)} \phi_j^{r,s,0}(K \cap E, \beta \cap E) \, \mu_k(dE) = \sum_{m=0}^{\lfloor \frac{s}{2} \rfloor} \sum_{i=0}^{1} d_{n,j,k}^{s,0,i,m} \, Q^{m-i} \phi_{n-k+j}^{r,s-2m,i}(K, \beta),$$

where

$$d_{n,j,k}^{s,0,i,m} = \frac{1}{(4\pi)^m m!} \frac{\binom{m}{i}}{\pi^i} \frac{\Gamma(\frac{n-k+j+1}{2}) \Gamma(\frac{k+1}{2})}{\Gamma(\frac{n+1}{2}) \Gamma(\frac{j+1}{2})}$$

$$\times \frac{\Gamma(\frac{n-k+j}{2} + 1)}{\Gamma(\frac{n-k+j+s}{2} + 1)} \frac{\Gamma(\frac{j+s}{2} - m + 1)}{\Gamma(\frac{j}{2} + 1)} \frac{\Gamma(\frac{n-k}{2} + m)}{\Gamma(\frac{n-k}{2})}.$$

In Theorem 4, we have $d_{n,j,k}^{s,0,1,0} = 0$ so that in fact the undefined tensor Q^{-1} does not appear.

For the special case $j = k - 1$, we deduce two more Crofton formulae. The first concerns the generalized tensorial curvature measures $\phi_{k-1}^{r,s,1}$.

Corollary 1 *Let $K \in \mathcal{K}^n$, $\beta \in \mathcal{B}(\mathbb{R}^n)$, and $k, r, s \in \mathbb{N}_0$ with $0 < k < n$. Then,*

$$\int_{A(n,k)} \phi_{k-1}^{r,s,1}(K \cap E, \beta \cap E) \, \mu_k(dE) = \sum_{m=0}^{\lfloor \frac{s}{2} \rfloor} \iota_{n,k}^{s,m} \, Q^m \phi_{n-1}^{r,s-2m,1}(K, \beta),$$

where

$$\iota_{n,k}^{s,m} := \frac{1}{(4\pi)^m m!} \frac{\Gamma(\frac{n}{2})\Gamma(\frac{k+s+1}{2} - m)\Gamma(\frac{n-k}{2} + m)}{\Gamma(\frac{n+s+1}{2})\Gamma(\frac{k}{2})\Gamma(\frac{n-k}{2})}.$$

Due to the easily verified relation

$$\phi_{n-1}^{r,s-2m,1} = \frac{2\pi}{n-1}\left(Q\phi_{n-1}^{r,s-2m,0} - 2\pi(s - 2m + 2)\phi_{n-1}^{r,s-2m+2,0}\right), \tag{7}$$

Corollary 1 can be transformed in such a way that only the tensorial curvature measures $\phi_{n-1}^{r,s-2m,0}$ are involved on the right-hand side of the preceding formula. This is presented in the following corollary.

Corollary 2 *Let $K \in \mathcal{K}^n$, $\beta \in \mathcal{B}(\mathbb{R}^n)$, and $k, r, s \in \mathbb{N}_0$ with $1 < k < n$. Then,*

$$\int_{A(n,k)} \phi_{k-1}^{r,s,1}(K \cap E, \beta \cap E)\, \mu_k(dE) = \sum_{m=0}^{\lfloor \frac{s}{2} \rfloor + 1} \lambda_{n,k}^{s,m} Q^m \phi_{n-1}^{r,s-2m+2,0}(K, \beta),$$

where

$$\lambda_{n,k}^{s,m} := \frac{\pi}{(n-1)(4\pi)^{m-1}m!} \frac{\Gamma(\frac{n}{2})\Gamma(\frac{k+s+1}{2} - m)\Gamma(\frac{n-k}{2} + m - 1)}{\Gamma(\frac{n+s+1}{2})\Gamma(\frac{k}{2})\Gamma(\frac{n-k}{2})}$$
$$\times \left(2m(\frac{k+s+1}{2} - m) - (s - 2m + 2)(\frac{n-k}{2} + m - 1)\right),$$

for $m \in \{1, \ldots, \lfloor \frac{s}{2} \rfloor\}$, and

$$\lambda_{n,k}^{s,0} := -\frac{4\pi^2(s+2)}{n-1} \frac{\Gamma(\frac{n}{2})\Gamma(\frac{k+s+1}{2})}{\Gamma(\frac{n+s+1}{2})\Gamma(\frac{k}{2})},$$

$$\lambda_{n,k}^{s,\lfloor \frac{s}{2} \rfloor + 1} := \frac{2\pi}{(n-1)(4\pi)^{\lfloor \frac{s}{2} \rfloor}(\lfloor \frac{s}{2} \rfloor)!} \frac{\Gamma(\frac{n}{2})\Gamma(\frac{k+s+1}{2} - \lfloor \frac{s}{2} \rfloor)\Gamma(\frac{n-k}{2} + \lfloor \frac{s}{2} \rfloor)}{\Gamma(\frac{n+s+1}{2})\Gamma(\frac{k}{2})\Gamma(\frac{n-k}{2})}.$$

The second special case concerns the tensorial curvature measures $\phi_{k-1}^{r,s,0}$. Although the result has been derived in a different way in our previous work [13, Theorem 4.12], we state it and derive it as a special case of the present more general approach.

Corollary 3 *Let $K \in \mathcal{K}^n$, $\beta \in \mathcal{B}(\mathbb{R}^n)$, and $k, r, s \in \mathbb{N}_0$ with $1 < k < n$. Then*

$$\int_{A(n,k)} \phi_{k-1}^{r,s,0}(K \cap E, \beta \cap E)\, \mu_k(dE) = \sum_{m=0}^{\lfloor \frac{s}{2} \rfloor} \kappa_{n,k}^{s,m} Q^m \phi_{n-1}^{r,s-2m,0}(K, \beta),$$

where

$$\kappa_{n,k}^{s,m} := \frac{k-1}{n-1} \frac{1}{(4\pi)^m m!} \frac{\Gamma(\frac{n}{2})\Gamma(\frac{k+s-1}{2} - m)\Gamma(\frac{n-k}{2} + m)}{\Gamma(\frac{n+s-1}{2})\Gamma(\frac{k}{2})\Gamma(\frac{n-k}{2})}$$

if $m \neq \frac{s-1}{2}$, and

$$\kappa_{n,k}^{s,\frac{s-1}{2}} := \frac{k(n+s-2)}{2(n-1)} \frac{1}{(4\pi)^{\frac{s-1}{2}} \frac{s-1}{2}!} \frac{\Gamma(\frac{n}{2})\Gamma(\frac{n-k+s-1}{2})}{\Gamma(\frac{n+s+1}{2})\Gamma(\frac{n-k}{2})}.$$

Finally, we state the remaining case where $k = 1$ (see also [13, Theorem 4.13]).

Corollary 4 *Let $K \in \mathcal{K}^n$, $\beta \in \mathcal{B}(\mathbb{R}^n)$, and $r, s \in \mathbb{N}_0$. Then*

$$\int_{A(n,1)} \phi_0^{r,s,0}(K \cap E, \beta \cap E)\,\mu_1(dE)$$

$$= \frac{\Gamma(\frac{s}{2} - \lfloor \frac{s}{2} \rfloor + 1)}{\sqrt{\pi}(4\pi)^{\lfloor \frac{s}{2} \rfloor} \lfloor \frac{s}{2} \rfloor!} \frac{\Gamma(\frac{n}{2})\Gamma(\frac{n+1}{2} + \lfloor \frac{s}{2} \rfloor)}{\Gamma(\frac{n+1}{2})\Gamma(\frac{n+s+1}{2})} Q^{\lfloor \frac{s}{2} \rfloor} \phi_{n-1}^{r,s-2\lfloor \frac{s}{2} \rfloor,0}(K, \beta).$$

Comparing Corollaries 3 and 4 to the corresponding results in [13], it should be observed that the normalization of the tensorial measures in [13] is different from the current normalization (although the measures are denoted in the same way).

4   The Proofs of the Main Results

In this section, we prove the Crofton formulae which have been stated in Sect. 3.

4.1   The Kinematic Formula for Generalized Tensorial Curvature Measures

The proof of the Crofton formulae uses the connection to the corresponding (more general) kinematic formulae. For the classical scalar-valued curvature measures this connection is well known. For easier reference, we state the required kinematic formula, which has recently been proved in [12, Theorem 1]. To state the result, we write G_n for the rigid motion group of $\mathbb{R}^n$ and denote by μ the Haar measure on G_n with the usual normalization (see [13], [27, p. 586]).

Theorem 5 (Kinematic Formula [12]) *For $P, P' \in \mathscr{P}^n$, $\beta, \beta' \in \mathscr{B}(\mathbb{R}^n)$, $j, l, r, s \in \mathbb{N}_0$ with $j \le n$, and $l = 0$ if $j = 0$,*

$$\int_{G_n} \phi_j^{r,s,l}(P \cap gP', \beta \cap g\beta') \, \mu(dg)$$

$$= \sum_{p=j}^{n} \sum_{m=0}^{\lfloor \frac{s}{2} \rfloor} \sum_{i=0}^{m} d_{n,j,n-p+j}^{s,l,i,m} \, Q^{m-i} \phi_p^{r,s-2m,l+i}(P, \beta) \phi_{n-p+j}(P', \beta'),$$

where

$$d_{n,j,n-p+j}^{s,l,i,m} = \frac{(-1)^i}{(4\pi)^m m!} \frac{\binom{m}{i}}{\pi^i} \frac{(i+l-2)!}{(l-2)!} \frac{\Gamma(\frac{n-p+j+1}{2})\Gamma(\frac{p+1}{2})}{\Gamma(\frac{n+1}{2})\Gamma(\frac{j+1}{2})}$$

$$\times \frac{\Gamma(\frac{p}{2} + 1)}{\Gamma(\frac{p+s}{2} + 1)} \frac{\Gamma(\frac{j+s}{2} - m + 1)}{\Gamma(\frac{j}{2} + 1)} \frac{\Gamma(\frac{p-j}{2} + m)}{\Gamma(\frac{p-j}{2})}.$$

In the formulation of Theorem 5, we changed the order of the coefficients slightly as compared to the original work (see [12, Theorem 1]), as we have $d_{n,j,n-p+j}^{s,l,i,m} = c_{n,j}^{s,l,i,m}$. This is done in order to shorten the representation of the Crofton formulae. Furthermore, since $\phi_n^{r,\tilde{s},l}$ vanishes for $\tilde{s} \ne 0$ and the functionals $Q^{\frac{s}{2}-i}\phi_n^{r,0,l+i}$, $i \in \{0, \ldots, \frac{s}{2}\}$, can be combined, we can redefine

$$d_{n,j,j}^{s,l,i,m} := \mathbf{1}\{s \text{ even}, m = i = \tfrac{s}{2}\} \frac{1}{(2\pi)^s(\frac{s}{2})!} \frac{\Gamma(\frac{n-j+s}{2})}{\Gamma(\frac{n-j}{2})};$$

for further details see the remark after Theorem 1 in [12]. In particular, the ratio $(i + l - 2)!/(l - 2)!$ is interpreted as a ratio of Gamma functions for $l \in \{0, 1\}$ (see also the comments after Theorem 2).

4.2 The Proofs

We prove both, Theorems 1 and 2, at once using the kinematic formula for generalized tensorial curvature measures deduced in [12] and restated in the last section as Theorem 5.

Proof (Theorems 1 and 2) Let $P \in \mathscr{P}^n$ and $\beta \in \mathscr{B}(\mathbb{R}^n)$. First, we prove the identity

$$J := \int_{A(n,k)} \phi_j^{r,s,l}(P \cap E, \beta) \, \mu_k(dE) = \int_{G_n} \phi_j^{r,s,l}(P \cap gE_k, \beta \cap g\alpha) \, \mu(dg) \qquad (8)$$

for an arbitrary (but fixed) $E_k \in G(n, k)$ and $\alpha \in \mathcal{B}(E_k)$ with $\mathcal{H}^k(\alpha) = 1$. This is shown as follows. Using (3), we obtain

$$J = \int_{SO(n)} \int_{E_k^\perp} \int_{\mathbb{R}^n} \mathbf{1}_\beta(x) \, \phi_j^{r,s,l}(P \cap \rho(E_k + t_1), \mathrm{d}x) \, \mathcal{H}^{n-k}(\mathrm{d}t_1) \, \nu(\mathrm{d}\rho).$$

For $t_1 \in E_k^\perp$ and $x \in \rho(E_k + t_1)$ we have

$$x \in \rho(\alpha + t_1 + t_2) \Leftrightarrow t_2 \in -\alpha + \rho^{-1}x - t_1,$$

for all $t_2 \in E_k$. Moreover, $-\alpha + \rho^{-1}x - t_1 \subset E_k$, since $\alpha \subset E_k$ and $x \in \rho(E_k + t_1)$ yields $\rho^{-1}x - t_1 \in E_k$. Thus, we get

$$\mathcal{H}^k\left(\{t_2 \in E_k : x \in \rho(\alpha + t_1 + t_2)\}\right) = \mathcal{H}^k(-\alpha + \rho^{-1}x - t_1) = \mathcal{H}^k(\alpha) = 1,$$

and hence we have

$$J = \int_{SO(n)} \int_{E_k^\perp} \int_{\mathbb{R}^n} \mathbf{1}_\beta(x) \int_{E_k} \mathbf{1}\{x \in \rho(\alpha + t_1 + t_2)\} \, \mathcal{H}^k(\mathrm{d}t_2)$$

$$\times \phi_j^{r,s,l}(P \cap \rho(E_k + t_1), \mathrm{d}x) \, \mathcal{H}^{n-k}(\mathrm{d}t_1) \, \nu(\mathrm{d}\rho)$$

$$= \int_{SO(n)} \int_{E_k^\perp} \int_{E_k} \int_{\mathbb{R}^n} \mathbf{1}_{\beta \cap \rho(\alpha+t_1+t_2)}(x) \, \phi_j^{r,s,l}(P \cap \rho(E_k + t_1 + t_2), \mathrm{d}x)$$

$$\times \mathcal{H}^k(\mathrm{d}t_2) \, \mathcal{H}^{n-k}(\mathrm{d}t_1) \, \nu(\mathrm{d}\rho).$$

Finally, Fubini's theorem yields

$$J = \int_{SO(n)} \int_{\mathbb{R}^n} \phi_j^{r,s,l}(P \cap \rho(E_k + t), \beta \cap \rho(\alpha + t)) \, \mathcal{H}^n(\mathrm{d}t) \, \nu(\mathrm{d}\rho)$$

$$= \int_{G_n} \phi_j^{r,s,l}(P \cap gE_k, \beta \cap g\alpha) \, \mu(\mathrm{d}g),$$

which concludes the proof of (8).

Let $\alpha \in \mathcal{B}(\mathbb{R}^n)$ be compact with $\alpha \subset E_k$ and $\mathcal{H}^k(\alpha) = 1$. Then choose $P' \in \mathcal{P}^n$ with $P' \subset E_k$ and $\alpha \subset \mathrm{relint}\, P'$, such that the following holds, for all $g \in G_n$: If $g^{-1}P \cap \alpha \neq \emptyset$, then $g^{-1}P \cap E_k \subset P'$. Hence, if $P \cap g\alpha \neq \emptyset$, then $P \cap gE_k = P \cap gP'$. Thus we obtain

$$J = \int_{G_n} \phi_j^{r,s,l}(P \cap gP', \beta \cap g\alpha) \, \mu(\mathrm{d}g),$$

and thus, by Theorem 5

$$J = \sum_{p=j}^{n} \sum_{m=0}^{\lfloor \frac{s}{2} \rfloor} \sum_{i=0}^{m} d_{n,j,n-p+j}^{s,l,i,m} Q^{m-i} \phi_p^{r,s-2m,l+i}(P,\beta) \phi_{n-p+j}(P',\alpha).$$

Hence, if $k = j$ we get

$$J = \mathbf{1}\{s \text{ even}\} \frac{1}{(2\pi)^s \left(\frac{s}{2}\right)!} \frac{\Gamma(\frac{n-k+s}{2})}{\Gamma(\frac{n-k}{2})} \phi_n^{r,0,\frac{s}{2}+l}(P,\beta) \underbrace{\phi_k(P',\alpha)}_{=\mathcal{H}^k(\alpha)=1}$$

$$= \mathbf{1}\{s \text{ even}\} \frac{1}{(2\pi)^s \frac{s}{2}!} \frac{\Gamma(\frac{n-k+s}{2})}{\Gamma(\frac{n-k}{2})} \phi_n^{r,0,\frac{s}{2}+l}(P,\beta),$$

and for $j < k$ we get

$$J = \sum_{m=0}^{\lfloor \frac{s}{2} \rfloor} \sum_{i=0}^{m} d_{n,j,k}^{s,l,i,m} Q^{m-i} \phi_{n-k+j}^{r,s-2m,l+i}(P,\beta) \underbrace{\phi_k(P',\alpha)}_{=\mathcal{H}^k(\alpha)=1}$$

$$= \sum_{m=0}^{\lfloor \frac{s}{2} \rfloor} \sum_{i=0}^{m} d_{n,j,k}^{s,l,i,m} Q^{m-i} \phi_{n-k+j}^{r,s-2m,l+i}(P,\beta),$$

since $\phi_q(P',\alpha) = 0$ for $q \neq k$.   □

Next, we prove Corollaries 1 and 2, which are derived from Theorem 3. The first follows immediately, whereas the second subsequently is obtained by an application of (7).

Proof (Corollaries 1 and 2) In both cases, we denote the integral we are interested in by I. First, we consider Corollary 1 and hence $l = 1$. In this case, Theorem 3 yields

$$I = \sum_{m=0}^{\lfloor \frac{s}{2} \rfloor} \iota_{n,k}^{s,m} Q^m \phi_{n-1}^{r,s-2m,1}(K,\beta),$$

where

$$\iota_{n,k}^{s,m} := d_{n,k-1,k}^{s,1,0,m} = \frac{1}{(4\pi)^m m!} \frac{\Gamma(\frac{n}{2})\Gamma(\frac{k+s+1}{2} - m)\Gamma(\frac{n-k}{2} + m)}{\Gamma(\frac{n+s+1}{2})\Gamma(\frac{k}{2})\Gamma(\frac{n-k}{2})},$$

which already proves Corollary 1.

Next, we turn to the proof of Corollary 2. From Corollary 1 and (7), we conclude that

$$I = \frac{2\pi}{n-1} \sum_{m=0}^{\lfloor \frac{s}{2} \rfloor} \iota_{n,k}^{s,m} Q^{m+1} \phi_{n-1}^{r,s-2m,0}(K,\beta) - 2\pi(s-2m+2)\iota_{n,k}^{s,m} Q^m \phi_{n-1}^{r,s-2m+2,0}(K,\beta)$$

$$= \frac{2\pi}{n-1} \sum_{m=1}^{\lfloor \frac{s}{2} \rfloor+1} \iota_{n,k}^{s,m-1} Q^m \phi_{n-1}^{r,s-2m+2,0}(K,\beta)$$

$$- \frac{2\pi}{n-1} \sum_{m=0}^{\lfloor \frac{s}{2} \rfloor} 2\pi(s-2m+2)\iota_{n,k}^{s,m} Q^m \phi_{n-1}^{r,s-2m+2,0}(K,\beta)$$

$$= \sum_{m=1}^{\lfloor \frac{s}{2} \rfloor} \frac{2\pi}{n-1} \left(\iota_{n,k}^{s,m-1} - 2\pi(s-2m+2)\iota_{n,k}^{s,m} \right) Q^m \phi_{n-1}^{r,s-2m+2,0}(K,\beta)$$

$$+ \frac{2\pi}{n-1} \iota_{n,k}^{s,\lfloor \frac{s}{2} \rfloor} Q^{\lfloor \frac{s}{2} \rfloor+1} \phi_{n-1}^{r,s-2\lfloor \frac{s}{2} \rfloor,0}(K,\beta) - \frac{4\pi^2(s+2)}{n-1} \iota_{n,k}^{s,0} \phi_{n-1}^{r,s+2,0}(K,\beta).$$

Denoting the coefficients by $\lambda_{n,k}^{s,m}$, we obtain for $m \in \{1, \ldots, \lfloor \frac{s}{2} \rfloor\}$

$$\lambda_{n,k}^{s,m} = \frac{\pi}{(n-1)(4\pi)^{m-1}m!} \frac{\Gamma(\frac{n}{2})\Gamma(\frac{k+s+1}{2} - m)\Gamma(\frac{n-k}{2} + m - 1)}{\Gamma(\frac{n+s+1}{2})\Gamma(\frac{k}{2})\Gamma(\frac{n-k}{2})}$$

$$\times \left(2m(\frac{k+s+1}{2} - m) - (s-2m+2)(\frac{n-k}{2} + m - 1) \right),$$

and

$$\lambda_{n,k}^{s,0} = -\frac{4\pi^2(s+2)}{n-1} \iota_{n,k}^{s,0} = -\frac{4\pi^2(s+2)}{n-1} \frac{\Gamma(\frac{n}{2})\Gamma(\frac{k+s+1}{2})}{\Gamma(\frac{n+s+1}{2})\Gamma(\frac{k}{2})},$$

$$\lambda_{n,k}^{s,\lfloor \frac{s}{2} \rfloor+1} = \frac{2\pi}{n-1} \iota_{n,k}^{s,\lfloor \frac{s}{2} \rfloor}$$

$$= \frac{2\pi}{(n-1)(4\pi)^{\lfloor \frac{s}{2} \rfloor}(\lfloor \frac{s}{2} \rfloor)!} \frac{\Gamma(\frac{n}{2})\Gamma(\frac{k+s+1}{2} - \lfloor \frac{s}{2} \rfloor)\Gamma(\frac{n-k}{2} + \lfloor \frac{s}{2} \rfloor)}{\Gamma(\frac{n+s+1}{2})\Gamma(\frac{k}{2})\Gamma(\frac{n-k}{2})},$$

where $\lambda_{n,k}^{s,0}$ is defined according to the general definition, but $\lambda_{n,k}^{s,\lfloor \frac{s}{2} \rfloor+1}$ differs slightly for odd s.   □

Finally, we prove Corollaries 3 and 4, which are derived from Theorem 4.

Proof (Corollaries 3 and 4) We denote the integral we are interested in by I and establish both corollaries simultaneously. Theorem 4 yields

$$I = \sum_{m=0}^{\lfloor \frac{s}{2} \rfloor} d_{n,k-1,k}^{s,0,0,m} Q^m \phi_{n-1}^{r,s-2m,0}(K,\beta) + \sum_{m=1}^{\lfloor \frac{s}{2} \rfloor} d_{n,k-1,k}^{s,0,1,m} Q^{m-1} \phi_{n-1}^{r,s-2m,1}(K,\beta),$$

where

$$d_{n,k-1,k}^{s,0,i,m} := \frac{1}{4^m(m-i)!} \frac{1}{\pi^{i+m}} \frac{\Gamma(\frac{n}{2})}{\Gamma(\frac{n+s+1}{2})\Gamma(\frac{k}{2})\Gamma(\frac{n-k}{2})} \Gamma(\tfrac{k+s+1}{2} - m)\Gamma(\tfrac{n-k}{2} + m).$$

From (7) we obtain that

$$I = \sum_{m=0}^{\lfloor \frac{s}{2} \rfloor} d_{n,k-1,k}^{s,0,0,m} Q^m \phi_{n-1}^{r,s-2m,0}(K,\beta) + \frac{2\pi}{n-1} \sum_{m=1}^{\lfloor \frac{s}{2} \rfloor} d_{n,k-1,k}^{s,0,1,m} Q^m \phi_{n-1}^{r,s-2m,0}(K,\beta)$$

$$- \frac{4\pi^2}{n-1} \sum_{m=1}^{\lfloor \frac{s}{2} \rfloor} d_{n,k-1,k}^{s,0,1,m}(s - 2m + 2) Q^{m-1} \phi_{n-1}^{r,s-2m+2,0}(K,\beta),$$

where we used that $d_{n,k-1,k}^{s,0,1,0} = 0$. This can be rewritten in the form

$$I = \sum_{m=0}^{\lfloor \frac{s}{2} \rfloor} \left(d_{n,k-1,k}^{s,0,0,m} + \frac{2\pi}{n-1} d_{n,k-1,k}^{s,0,1,m} \right) Q^m \phi_{n-1}^{r,s-2m,0}(K,\beta)$$

$$- \frac{4\pi^2}{n-1} \sum_{m=0}^{\lfloor \frac{s}{2} \rfloor - 1} d_{n,k-1,k}^{s,0,1,m+1}(s - 2m) Q^m \phi_{n-1}^{r,s-2m,0}(K,\beta)$$

$$= \sum_{m=0}^{\lfloor \frac{s}{2} \rfloor - 1} \left(d_{n,k-1,k}^{s,0,0,m} + \frac{2\pi}{n-1} d_{n,k-1,k}^{s,0,1,m} - \frac{4\pi^2(s-2m)}{n-1} d_{n,k-1,k}^{s,0,1,m+1} \right) Q^m \phi_{n-1}^{r,s-2m,0}(K,\beta)$$

$$+ \left(d_{n,k-1,k}^{s,0,0,\lfloor \frac{s}{2} \rfloor} + \frac{2\pi}{n-1} d_{n,k-1,k}^{s,0,1,\lfloor \frac{s}{2} \rfloor} \right) Q^{\lfloor \frac{s}{2} \rfloor} \phi_{n-1}^{r,s-2\lfloor \frac{s}{2} \rfloor,0}(K,\beta).$$

Denoting the corresponding coefficients of the summand $Q^m \phi_{n-1}^{r,s-2m,0}(K,\beta)$ by $\kappa_{n,k}^{s,m}$, we obtain

$$\kappa_{n,k}^{s,m} = \left(1 + \frac{2m}{n-1} \right) \frac{1}{(4\pi)^m m!} \frac{\Gamma(\frac{n}{2})\Gamma(\frac{k+s+1}{2} - m)\Gamma(\frac{n-k}{2} + m)}{\Gamma(\frac{n+s+1}{2})\Gamma(\frac{k}{2})\Gamma(\frac{n-k}{2})}$$

$$- \frac{s-2m}{n-1} \frac{1}{(4\pi)^m m!} \frac{\Gamma(\frac{n}{2})\Gamma(\frac{k+s-1}{2} - m)\Gamma(\frac{n-k}{2} + m + 1)}{\Gamma(\frac{n+s+1}{2})\Gamma(\frac{k}{2})\Gamma(\frac{n-k}{2})}$$

$$= \frac{1}{(4\pi)^m m!} \frac{\Gamma(\frac{n}{2})\Gamma(\frac{k+s-1}{2} - m)\Gamma(\frac{n-k}{2} + m)}{\Gamma(\frac{n+s+1}{2})\Gamma(\frac{k}{2})\Gamma(\frac{n-k}{2})}$$

$$\times \underbrace{\left(\frac{n+2m-1}{n-1}(\tfrac{k+s-1}{2} - m) - \frac{s-2m}{n-1}(\tfrac{n-k}{2} + m) \right)}_{= \frac{k-1}{n-1} \frac{n+s-1}{2}}$$

$$= \frac{k-1}{n-1} \frac{1}{(4\pi)^m m!} \frac{\Gamma(\frac{n}{2})\Gamma(\frac{k+s-1}{2} - m)\Gamma(\frac{n-k}{2} + m)}{\Gamma(\frac{n+s-1}{2})\Gamma(\frac{k}{2})\Gamma(\frac{n-k}{2})}, \tag{9}$$

for $m \in \{0, \ldots, \lfloor \frac{s}{2} \rfloor - 1\}$. For $k = 1$, we immediately get $\kappa_{n,1}^{s,m} = 0$ in these cases. Furthermore, we have

$$\kappa_{n,k}^{s,\lfloor \frac{s}{2} \rfloor} = \left(1 + \frac{2\lfloor \frac{s}{2} \rfloor}{n-1}\right) \frac{1}{(4\pi)^{\lfloor \frac{s}{2} \rfloor} \lfloor \frac{s}{2} \rfloor!} \frac{\Gamma(\frac{n}{2})\Gamma(\frac{k+s+1}{2} - \lfloor \frac{s}{2} \rfloor)\Gamma(\frac{n-k}{2} + \lfloor \frac{s}{2} \rfloor)}{\Gamma(\frac{n+s+1}{2})\Gamma(\frac{k}{2})\Gamma(\frac{n-k}{2})}.$$

If s is even and $k > 1$, this coincides with (9) for $m = \frac{s}{2}$. If s is odd, we have

$$\kappa_{n,k}^{s,\frac{s-1}{2}} = \frac{k(n+s-2)}{2(n-1)} \frac{1}{(4\pi)^{\frac{s-1}{2}} \frac{s-1}{2}!} \frac{\Gamma(\frac{n}{2})\Gamma(\frac{n-k+s-1}{2})}{\Gamma(\frac{n+s+1}{2})\Gamma(\frac{n-k}{2})}$$

and thus the assertion of Corollary 3. For $k = 1$, we obtain

$$\kappa_{n,1}^{s,\lfloor \frac{s}{2} \rfloor} = \frac{\Gamma(\frac{s}{2} - \lfloor \frac{s}{2} \rfloor + 1)}{\sqrt{\pi}(4\pi)^{\lfloor \frac{s}{2} \rfloor} \lfloor \frac{s}{2} \rfloor!} \frac{\Gamma(\frac{n}{2})\Gamma(\frac{n+1}{2} + \lfloor \frac{s}{2} \rfloor)}{\Gamma(\frac{n+1}{2})\Gamma(\frac{n+s+1}{2})}$$

and thus the assertion of Corollary 4.   □

5 Linear Independence of the Generalized Tensorial Curvature Measures

In this section, we show that the generalized tensorial curvature measures multiplied with powers of the metric tensor are linearly independent. The proof of this result follows the argument for Theorem 3.1 in [9].

Theorem 6 *For $p \in \mathbb{N}_0$, the tensorial measure valued valuations*

$$Q^m \phi_j^{r,s,l} : \mathscr{P}^n \times \mathscr{B}(\mathbb{R}^n) \to \mathbb{T}^p$$

with $m, r, s, l \in \mathbb{N}_0$ and $j \in \{0, \ldots, n\}$, where $2m + 2l + r + s = p$, but with $l = 0$ if $j \in \{0, n-1\}$ and with $s = l = 0$ if $j = n$, are linearly independent.

In particular, Theorem 6 shows that the Crofton formulae, which we stated in Sect. 3 and proved in Sect. 4, cannot be simplified further, as there are no more linear dependencies between the appearing functionals. A corresponding statement holds for the results from our preceding work [12].

Proof Suppose that

$$\sum_{\substack{j,m,r,s,l \\ 2m+2l+r+s=p}} a_{j,m,r,s,l}^{(0)} Q^m \phi_j^{r,s,l} = 0 \qquad (10)$$

holds for some $a^{(0)}_{j,m,r,s,l} \in \mathbb{R}$, where $a^{(0)}_{0,m,r,s,l} = a^{(0)}_{n-1,m,r,s,l} = 0$ if $l \neq 0$ and $a^{(0)}_{n,m,r,s,l} = 0$ if $s \neq 0$ or $l \neq 0$. In the proof we will replace the constants $a^{(0)}_{j,m,r,s,l}$ by new constants $a^{(1)}_{j,m,r,s,l}$ without keeping track of the precise relations, since it will be sufficient to know that $a^{(0)}_{j,m,r,s,l} = 0$ if and only if $a^{(1)}_{j,m,r,s,l} = 0$.

For a fixed $j \in \{0, \ldots, n\}$, let $P \in \mathscr{P}^n$ with $\operatorname{int} P \neq \emptyset$, $F \in \mathscr{F}_j(P)$, and $\beta \in \mathscr{B}(\operatorname{relint} F)$. Then, if $j < n$ we obtain for the generalized tensorial curvature measures

$$\phi_j^{r,s,l}(P,\beta) = c_{n,j,r,s,l} \sum_{G \in \mathscr{F}_j(P)} Q(G)^l \int_{G \cap \beta} x^r \, \mathscr{H}^j(\mathrm{d}x) \int_{N(P,G) \cap \mathbb{S}^{n-1}} u^s \, \mathscr{H}^{n-j-1}(\mathrm{d}u)$$

$$= c_{n,j,r,s,l} Q(F)^l \int_\beta x^r \, \mathscr{H}^j(\mathrm{d}x) \int_{N(P,F) \cap \mathbb{S}^{n-1}} u^s \, \mathscr{H}^{n-j-1}(\mathrm{d}u),$$

where $c_{n,j,r,s,l} > 0$ is a constant, and $\phi_k^{r,s,l}(P,\beta) = 0$ for $k \neq j$. Moreover, we have

$$\phi_n^{r,0,0}(P,\beta) = \frac{1}{r!} \int_\beta x^r \, \mathscr{H}^n(\mathrm{d}x).$$

Hence, from (10) it follows that

$$\sum_{\substack{m,r,s,l \\ 2m+2l+r+s=p}} a^{(1)}_{j,m,r,s,l} Q^m Q(F)^l \int_\beta x^r \, \mathscr{H}^j(\mathrm{d}x) \int_{N(P,F) \cap \mathbb{S}^{n-1}} u^s \, \mathscr{H}^{n-j-1}(\mathrm{d}u) = 0,$$

where for $j = n$ the spherical integral is omitted (also in the following).

We may assume that $\int_\beta x^r \, \mathscr{H}^j(\mathrm{d}x) \neq 0$ (otherwise, we consider a translate of P and β). If we repeat the above calculations with multiples of P and β, a comparison of the degrees of homogeneity yields, for every $r \in \mathbb{N}_0$, that

$$\sum_{\substack{m,s,l \\ 2m+2l+s=p-r}} a^{(1)}_{j,m,r,s,l} Q^m Q(F)^l \int_\beta x^r \, \mathscr{H}^j(\mathrm{d}x) \int_{N(P,F) \cap \mathbb{S}^{n-1}} u^s \, \mathscr{H}^{n-j-1}(\mathrm{d}u) = 0.$$

Hence, due to the lack of zero divisors in the tensor algebra $\mathbb{T}$, we obtain

$$\sum_{\substack{m,s,l \\ 2m+2l+s=p-r}} a^{(1)}_{j,m,r,s,l} Q^m Q(F)^l \int_{N(P,F) \cap \mathbb{S}^{n-1}} u^s \, \mathscr{H}^{n-j-1}(\mathrm{d}u) = 0. \qquad (11)$$

This shows that, in the case of $j = n$ (where the spherical integral is omitted), we have $a^{(1)}_{n,m,r,s,l} = 0$ also for $s = l = 0$. Hence, in the following we may assume that $j < n$.

Let $L \in G(n,j)$, $j < n$, and $u_0 \in L^\perp \cap \mathbb{S}^{n-1}$. For $j \leq n-2$, let $u_0, u_1, \ldots, u_{n-j-1}$ be an orthonormal basis of $L^\perp$. In this case, we define the (pointed) polyhedral cone $C(u_0, \tau) := \operatorname{pos}\{u_0 \pm \tau u_1, \ldots, u_0 \pm \tau u_{n-j-1}\} \subset L^\perp$ for $\tau \in (0,1)$. Then, for any $v \in C(u_0, \tau) \cap \mathbb{S}^{n-1}$, we have $\langle v, u_0 \rangle \geq 1/\sqrt{1+\tau^2}$, and therefore $\|u_0 - v\| \leq \sqrt{2}\tau$. In fact, any $v \in C(u_0, \tau) \cap \mathbb{S}^{n-1}$ can be written as $v = \frac{x}{\|x\|}$, where $x \in \operatorname{conv}\{v_1^\pm, \ldots, v_{n-j-1}^\pm\}$ with $v_i^\pm = \frac{u_0 \pm \tau u_i}{\|u_0 \pm \tau u_i\|} = \frac{u_0 \pm \tau u_i}{\sqrt{1+\tau^2}} \in \mathbb{S}^{n-1}$, $i \in \{1, \ldots, n-j-1\}$. Thus we have $x = \sum \lambda_i^\epsilon v_i^\epsilon$, where we sum over all $i \in \{1, \ldots, n-j-1\}$ and all $\epsilon \in \{+,-\}$, with $\sum \lambda_i^\epsilon = 1$ and $\lambda_i^\epsilon \geq 0$, $i \in \{1, \ldots, n-j-1\}$, $\epsilon \in \{+,-\}$. This yields

$$\langle v, u_0 \rangle = \frac{1}{\|x\|} \langle \sum \lambda_i^\epsilon v_i^\epsilon, u_0 \rangle = \frac{1}{\sqrt{1+\tau^2}\|x\|} \sum \lambda_i^\epsilon \geq \frac{1}{\sqrt{1+\tau^2}},$$

as $\|x\| \leq \sum \lambda_i^\epsilon \|v_i^\epsilon\| = 1$. This proves the assertion. For $j = n-1$ we simply put $C(u_0, \tau) := \operatorname{pos}\{u_0\}$.

Let $C(u_0, \tau)^\circ$ denote the polar cone of $C(u_0, \tau)$. Then $P := C(u_0, \tau)^\circ \cap [-1,1]^n \in \mathscr{P}^n$ and $F := L \cap [-1,1]^n \in \mathscr{F}_j(P)$ satisfy $N(P,F) = N(P,0) = C(u_0, \tau)$. With these choices, (11) turns into

$$\sum_{\substack{m,s,l \\ 2m+2l+s=p-r}} a_{j,m,r,s,l}^{(1)} Q^m Q(L)^l \int_{C(u_0,\tau)\cap\mathbb{S}^{n-1}} u^s \, \mathscr{H}^{n-j-1}(\mathrm{d}u) = 0. \tag{12}$$

Dividing (12) by $\mathscr{H}^{n-j-1}(C(u_0, \tau) \cap \mathbb{S}^{n-1})$ and passing to the limit as $\tau \to 0$, we get

$$\sum_{\substack{m,s,l \\ 2m+2l+s=p-r}} a_{j,m,r,s,l}^{(1)} Q^m Q(L)^l u_0^s = 0$$

for any $u_0 \in L^\perp \cap \mathbb{S}^{n-1}$. Here we use that

$$\left| \mathscr{H}^{n-j-1}(C(u_0,\tau)\cap\mathbb{S}^{n-1})^{-1} \int_{C(u_0,\tau)\cap\mathbb{S}^{n-1}} u^s \, \mathscr{H}^{n-j-1}(\mathrm{d}u) - u_0^s \right|$$

$$\leq \max\{|u^s - u_0^s| : u \in C(u_0,\tau)\cap\mathbb{S}^{n-1}\} \to 0$$

as $\tau \to 0$.

The rest of the proof follows similarly as in the proof of [9, Theorem 3.1]. □

Acknowledgements The authors were supported in part by DFG grants FOR 1548.

References

1. S. Alesker, Description of continuous isometry covariant valuations on convex sets. Geom. Dedicata **74**, 241–248 (1999)
2. E. Artin, *The Gamma Function* (Holt, Rinehart and Winston, New York, 1964)
3. A. Bernig, D. Hug, Kinematic formulas for tensor valuations. J. Reine Angew. Math. (2015). arXiv:1402.2750v2. https://doi.org/10.1515/crelle-2015-002
4. H. Federer, Curvature measures. Trans. Amer. Math. Soc. **93**, 418–491 (1959)
5. S. Glasauer, Integralgeometrie konvexer Körper im sphärischen Raum. Dissertation. Albert-Ludwigs-Universität Freiburg, Freiburg (1995)
6. S. Glasauer, A generalization of intersection formulae of integral geometry. Geom. Dedicata **68**, 101–121 (1997)
7. H. Hadwiger, Additive Funktionale k-dimensionaler Eikörper I. Arch. Math. **3**, 470–478 (1952)
8. H. Hadwiger, R. Schneider, Vektorielle Integralgeometrie. Elem. Math. **26**, 49–57 (1971)
9. D. Hug, R. Schneider, Local tensor valuations. Geom. Funct. Anal. **24**, 1516–1564 (2014)
10. D. Hug, R. Schneider, SO(n) covariant local tensor valuations on polytopes. Mich. Math. J. **66**, 637–659 (2017)
11. D. Hug, R. Schneider, Rotation covariant local tensor valuations on convex bodies. Commun. Contemp. Math. **19**, 1650061, 31 pp. (2017). https://doi.org/10.1142/S0219199716500619
12. D. Hug, J.A. Weis, Kinematic formulae for tensorial curvature measures. Ann. Mat. Pura Appl. arXiv: 1612.08427 (2016)
13. D. Hug, J.A. Weis, Crofton formulae for tensor-valued curvature measures, in *Tensor Valuations and their Applications in Stochastic Geometry and Imaging*, ed. by M. Kiderlen, E.B. Vedel Jensen. Lecture Notes in Mathematics, vol. 2177 (Springer, Berlin, 2017), pp. 111–156. https://doi.org/10.1007/978-3-319-51951-7_5
14. D. Hug, J.A. Weis, Integral formulae for Minkowski tensors. arXiv: 1712.09699 (2017)
15. D. Hug, R. Schneider, R. Schuster, Integral geometry of tensor valuations. Adv. Appl. Math. **41**, 482–509 (2008)
16. D. Hug, R. Schneider, R. Schuster, The space of isometry covariant tensor valuations. Algebra i Analiz **19**, 194–224 (2007); St. Petersburg Math. J. **19**, 137–158 (2008)
17. M. Kiderlen, E.B. Vedel Jensen, *Tensor Valuations and their Applications in Stochastic Geometry and Imaging*. Lecture Notes in Mathematics, vol. 2177 (Springer, Berlin, 2017)
18. P. McMullen, Isometry covariant valuations on convex bodies. Rend. Circ. Mat. Palermo (2), Suppl. **50**, 259–271 (1997)
19. K.R. Mecke, Additivity, convexity, and beyond: applications of Minkowski functionals in statistical physics, in *Statistical Physics and Spatial Statistics*, ed. by K.R. Mecke, D. Stoyan. Lecture Notes in Physics, vol. 554 (Springer, Berlin, 2000)
20. M. Saienko, Tensor-valued valuations and curvature measures in Euclidean spaces. PhD Thesis, University of Frankfurt (2016)
21. R. Schneider, Krümmungsschwerpunkte konvexer Körper. I. Abh. Math. Sem. Univ. Hamburg **37**, 112–132 (1972)
22. R. Schneider, Krümmungsschwerpunkte konvexer Körper. II. Abh. Math. Sem. Univ. Hamburg **37**, 204–217 (1972)
23. R. Schneider, Kinematische Berührmaße für konvexe Körper. Abh. Math. Sem. Univ. Hamburg **44**, 12–23 (1975)
24. R. Schneider, Curvature measures of convex bodies. Ann. Mat. Pura Appl. **116**, 101–134 (1978)
25. R. Schneider, Local tensor valuations on convex polytopes. Monatsh. Math. **171**, 459–479 (2013)

26. R. Schneider, *Convex Bodies: The Brunn-Minkowski Theory*. Encyclopedia of Mathematics and Its Applications, vol. 151 (Cambridge University Press, Cambridge, 2014)
27. R. Schneider, W. Weil, *Stochastic and Integral Geometry* (Springer, Berlin, 2008)
28. G.E. Schröder-Turk et al., Minkowski tensor shape analysis of cellular, granular and porous structures. Adv. Mat. **23**, 2535–2553 (2011)
29. A.M. Svane, E.B. Vedel Jensen, Rotational Crofton formulae for Minkowski tensors and some affine counterparts. Adv. Appl. Math. **91**, 44–75 (2017)

Extensions of Reverse Volume Difference Inequalities

Alexander Koldobsky and Denghui Wu

Abstract Volume difference inequalities are designed to estimate the difference between volumes of two bodies in terms of the maximal or minimal difference between areas of sections of these bodies. In this note we extend two such inequalities established in Koldobsky (Adv Math 283:473–488, 2015) and Giannopoulos and Koldobsky (Trans Am Math Soc, https://doi.org/10.1090/tran/7173, to appear) from the hyperplane case to the case of sections of arbitrary dimensions.

1 Introduction

The following volume difference inequality was proved in [10] for $k = 1$, and in [13] for $1 < k < n$. Let $1 \leq k < n$, let K be a generalized k-intersection body in $\mathbb{R}^n$ (we write $K \in \mathcal{BP}_k^n$; see definition in Sect. 2), and let L be any origin-symmetric star body L in $\mathbb{R}^n$ so that

$$\max_{F \in G_{n,n-k}} (|K \cap F| - |L \cap F|) > 0,$$

where $G_{n,n-k}$ is the Grassmannian of $(n-k)$-dimensional subspaces of $\mathbb{R}^n$, and $|\cdot|$ stands for volume of appropriate dimension. Then

$$|K|^{\frac{n-k}{n}} - |L|^{\frac{n-k}{n}} \leq c_{n,k}^k \max_{F \in G_{n,n-k}} (|K \cap F| - |L \cap F|), \qquad (1.1)$$

A. Koldobsky (✉)
Department of Mathematics, University of Missouri, Columbia, MO, USA
e-mail: koldobskiya@missouri.edu

D. Wu
School of Mathematics and Statistics, Southwest University, Chongqing, China

Department of Mathematics, University of Missouri, Columbia, MO, USA
e-mail: wudenghui66@163.com

© Springer International Publishing AG 2018
G. Bianchi et al. (eds.), *Analytic Aspects of Convexity*, Springer INdAM Series 25,
https://doi.org/10.1007/978-3-319-71834-7_4

where

$$c_{n,k}^k = \frac{\omega_n^{\frac{n-k}{n}}}{\omega_{n-k}},$$

and $\omega_n = |B_2^n|$ is the volume of the unit Euclidean ball in $\mathbb{R}^n$. Note that $c_{n,k} \in \left(1/\sqrt{e}, 1\right)$; see, for example, [12, Lemma 2.1].

A volume difference inequality of a different kind was proved in [3] for $k = 1$, and in [4] for $1 \le k < n$. Let $1 \le k < n$, and let L be a generalized k-intersection body in $\mathbb{R}^n$. Then, for any origin-symmetric star body K in $\mathbb{R}^n$ such that $|L \cap F| \le |K \cap F|$ for all $F \in G_{n,n-k}$,

$$|K|^{\frac{n-k}{n}} - |L|^{\frac{n-k}{n}} \ge c^k \frac{1}{(\sqrt{n}M(\bar{L}))^k} \min_{F \in G_{n,n-k}} (|K \cap F| - |L \cap F|), \tag{1.2}$$

where $c > 0$ is an absolute constant, $\bar{L} = L/|L|^{1/n}$, $M(L) = \int_{S^{n-1}} ||\theta|| d\sigma(\theta)$, and σ is the normalized Lebesgue measure on the sphere. Note that when L is convex, has volume 1 and is in the minimal mean width position, then we have

$$\frac{1}{M(L)} \ge c \frac{\sqrt{n}}{\log(e+n)}, \tag{1.3}$$

where c is an absolute constant; see [1].

Volume difference inequalities estimate the error in computations of volume of a body out of areas of its sections. They are closely related to Bourgain's slicing problem and to the Busemann-Petty problem; see [4] for details. We also refer to [4] for different extensions of (1.1) and (1.2) to arbitrary star and convex bodies and to arbitrary measures in place of volume. The paper [4] also provides volume difference inequalities for projections.

In the hyperplane case ($k = 1$), inequalities going in the opposite directions to (1.1) and (1.2) were proved in [4, 11]. It was shown in [4, Theorem 1.9] that for any $n \ge 5$ there exist origin-symmetric convex bodies K, L in $\mathbb{R}^n$ such that $L \subset K$ and

$$|K|^{\frac{n-1}{n}} - |L|^{\frac{n-1}{n}} > c \frac{1}{\sqrt{n}M(\bar{L})} \max_{\xi \in S^{n-1}} (|K \cap \xi^\perp| - |L \cap \xi^\perp|), \tag{1.4}$$

where $c > 0$ is an absolute constant, and $\xi^\perp$ is the central hyperplane in $\mathbb{R}^n$ perpendicular to ξ. The result in [11, Section 6] shows that for $n \ge 5$ there exist origin-symmetric convex bodies K, L in $\mathbb{R}^n$ such that $L \subset K$ and

$$|K|^{\frac{n-1}{n}} - |L|^{\frac{n-1}{n}} < c_{n,1} \min_{\xi \in S^{n-1}} (|K \cap \xi^\perp| - |L \cap \xi^\perp|). \tag{1.5}$$

We call (1.5) and (1.4) reverse volume difference inequalities. Combined with (1.3), these inequalities show that in the case $k = 1$ volume difference inequalities (1.1) and (1.2) are optimal up to a logarithmic term. In this note we extend the reverse volume difference inequalities to sections of codimensions $1 < k < n - 3$.

2 Preliminaries

Let $i \in \mathbb{N} \cup \{0\}$. Let $C^i(S^{n-1})$ be the space of all i-smooth functions on the unit sphere S^{n-1}, and let $C_e^i(S^{n-1})$ be the subspace of even functions in $C^i(S^{n-1})$. We denote by $C^i(G_{n,n-k})$ the space of all i-smooth functions on the Grassmann manifold $G_{n,n-k}$ of $(n - k)$-dimensional subspaces of $\mathbb{R}^n$.

A compact set $K \subset \mathbb{R}^n$ is called *star-shaped* if it contains the origin as its interior point, and $\alpha K \subset K$ for any $\alpha \in (0, 1)$. We say that $K \subset \mathbb{R}^n$ is a *star body* if it is star-shaped and its *Minkowski functional*

$$||x||_K := \min\{a \geq 0 : x \in aK\}$$

is a continuous function on $\mathbb{R}^n$. If K is an origin-symmetric convex body, then the Minkowski functional $|| \cdot ||_K$ is a norm on $\mathbb{R}^n$. We say that a star body K is *i-smooth*, $i \in \mathbb{N} \cup \{0\}$, if the restriction of $|| \cdot ||_K$ to the unit sphere S^{n-1} belongs to the class $C^i(S^{n-1})$. If $|| \cdot ||_K \in C^i(S^{n-1})$ for all $i \in \mathbb{N}$, then the body K is called *infinitely smooth*.

The *radial function* of a star body K is defined by

$$\rho_K(x) = ||x||_K^{-1}, \quad \text{for } x \in \mathbb{R}^n, x \neq 0.$$

If $x \in S^{n-1}$, then $\rho_K(x)$ is the radius of K in the direction of x.

Using polar coordinates, one can write volume of a star body K as

$$|K| = \frac{1}{n} \int_{S^{n-1}} ||\theta||_K^{-n} d\theta, \tag{2.1}$$

where $d\theta$ denotes the uniform measure on the sphere with density 1.

The main tool used in this paper is the Fourier transform of distribution; we refer the reader to [2] for details. Denote by $\mathcal{S}(\mathbb{R}^n)$ the *Schwartz space* of rapidly decreasing infinitely differentiable functions on $\mathbb{R}^n$ (also referred to as *test functions*), and by $\mathcal{S}'(\mathbb{R}^n)$ the space of *distributions* on $\mathbb{R}^n$, the space of continuous linear functionals on $\mathcal{S}(\mathbb{R}^n)$. The Fourier transform $\hat{f}$ of a distribution f is defined by $\langle \hat{f}, \varphi \rangle = \langle f, \hat{\varphi} \rangle$ for every test function φ. The Fourier transform is self-invertible up to a constant factor in the sense that $(\varphi^\wedge)^\wedge = (2\pi)^n \varphi$ for any even test function φ. A distribution f on $\mathbb{R}^n$ is called *even homogeneous of degree* $p \in \mathbb{R}$, if

$$\left\langle f(x), \varphi\left(\frac{x}{\alpha}\right) \right\rangle = |\alpha|^{n+p} \langle f, \varphi \rangle$$

for every test function φ and every $\alpha \in \mathbb{R}, \alpha \neq 0$. The Fourier transform of an even homogeneous distribution of degree p is an even homogeneous distribution of degree $-n - p$. We call a distribution f *positive definite*, if for every test function

$$\langle f(x), \varphi * \overline{\varphi(-x)} \rangle \geq 0.$$

By Schwartz's generalization of Bochner's theorem, a distribution f is positive definite if and only if $\hat{f}$ is a positive distribution, i.e. $\langle \hat{f}, \varphi \rangle \geq 0$ for every non-negative test function φ (see [9, Section 2.5] for details).

For an infinitely smooth star body K, the Fourier transform in the sense of distributions of $\|x\|_K^p$, $-n < p < 0$ is an infinitely smooth function g on S^{n-1} extended to a homogeneous function of the order $-n - p$ on the whole of $\mathbb{R}^n$; see [9, Lemma 3.16]. When we write $(\|x\|_K^p)^\wedge(\theta)$, we mean $g(\theta)$, for $\theta \in S^{n-1}$.

We use the following version of Parseval's formula on the sphere.

Proposition 2.1 ([9, Theorem 3.22]) *Let K and L be two infinitely smooth origin-symmetric star bodies in $\mathbb{R}^n$ and $0 < p < n$. Then*

$$\int_{S^{n-1}} (\|\cdot\|_K^{-p})^\wedge(\theta)(\|\cdot\|_L^{-n+p})^\wedge(\theta)d\theta = (2\pi)^n \int_{S^{n-1}} \|\theta\|_K^{-p}\|\theta\|_L^{-n+p}d\theta. \quad (2.2)$$

We also use a connection between the Fourier and Radon transforms.

Proposition 2.2 ([9, Theorem 3.24]) *Let $1 \leq k < n$, let ϕ be an even continuous integrable function on $\mathbb{R}^n$, and let H be an $(n - k)$-dimensional subspace of $\mathbb{R}^n$. Suppose that ϕ is integrable on all translations of H and that the Fourier transform $\hat{\phi}$ is integrable on $H^\perp$. Then*

$$(2\pi)^k \int_H \phi(x)dx = \int_{H^\perp} \hat{\phi}(x)dx. \quad (2.3)$$

We also need formulas expressing volume of central sections of star bodies in terms of the Fourier transform. It was proved in [7] (see also [9, Lemma 3.8]) that for every origin-symmetric star body K and for every $\xi \in S^{n-1}$

$$|K \cap \xi^\perp| = \frac{1}{\pi(n-1)}(\|\cdot\|_K^{-n+1})^\wedge(\xi).$$

A lower dimensional version of the latter formula is as follows.

Proposition 2.3 ([9, Theorem 3.25]) *Let $1 \leq k < n$, and let K be an origin-symmetric star body in $\mathbb{R}^n$ such that K is $(k - 1)$-smooth if k is odd and k-smooth if k is even. Suppose that H is an $(n - k)$-dimensional subspace of $\mathbb{R}^n$. Then*

$$|K \cap H| = \frac{1}{(2\pi)^k(n-k)} \int_{S^{n-1} \cap H^\perp} (\|\cdot\|_K^{-n+k})^\wedge(\theta)d\theta. \quad (2.4)$$

In particular, if $K = B_2^n$ is the unit Euclidean ball in $\mathbb{R}^n$, then for every $H \in G_{n,n-k}$ and every $\xi \in S^{n-1}$,

$$|B_2^n \cap H| = \frac{k\omega_k}{(2\pi)^k (n-k)} (||\cdot||_2^{-n+k})^\wedge(\xi), \qquad (2.5)$$

where

$$\omega_n = |B_2^n| = \frac{\pi^{n/2}}{\Gamma(1 + \frac{n}{2})}.$$

The concept of an intersection body was introduced by Lutwak [14] (for $k = 1$) and generalized to $k > 1$ by Zhang [15]. We formulate the definition in an equivalent form established by Goodey-Weil [5] and Grinberg-Zhang [6]. Let $1 \leq k \leq n - 1$. The class $\mathcal{BP}_k^n$ of *generalized k-intersection bodies* in $\mathbb{R}^n$ is the closure in the radial metric of radial k-sums of finite collections of origin-symmetric ellipsoids in $\mathbb{R}^n$. When $k = 1$, $\mathcal{BP}_1^n = \mathcal{I}_n$ is the original class of intersection bodies introduced by Lutwak.

Recall that the radial k-sum of star bodies K and L in $\mathbb{R}^n$ is a new star body $K +_k L$ whose radius in every direction $\xi \in S^{n-1}$ is given by

$$\rho_{K+_kL}^k(\xi) = \rho_K^k(\xi) + \rho_L^k(\xi), \qquad \forall \xi \in S^{n-1}.$$

The radial metric in the class of origin-symmetric star bodies is defined by

$$\rho(K,L) = \sup_{\xi \in S^{n-1}} |\rho_K(\xi) - \rho_L(\xi)|.$$

Another generalization of the concept of an intersection body was introduced in [8]. For an integer k, $1 \leq k < n$ and star bodies D, L in $\mathbb{R}^n$, we say that D is the k-intersection body of L if

$$|D \cap H^\perp| = |L \cap H|, \qquad \forall H \in Gr_{n-k}.$$

Taking the closure in the radial metric of the class of k-intersection bodies of star bodies, we define the class of k-intersection bodies $\mathcal{I}_k^n$. If $k = 1$, we get the original class of intersection bodies $\mathcal{I}_n$.

The following Fourier analytic characterization of k-intersection bodies was proved in [8].

Proposition 2.4 ([9, Theorem 4.8]) *Let $1 \leq k < n$. An origin-symmetric star body D in $\mathbb{R}^n$ is a k-intersection body if and only if the function $||\cdot||^{-k}$ represents a positive definite distribution on $\mathbb{R}^n$.*

By [9, Theorem 4.13, Lemma 4.10], for every $1 \leq k < n - 3$ there exist infinitely smooth convex bodies with positive curvature in $\mathbb{R}^n$ which are not k-intersection bodies. We will use these bodies to prove the reverse volume difference inequalities for $1 \leq k < n-3$. Note that by [9, Corollary 4.9] every origin-symmetric convex body in $\mathbb{R}^n$ is a k-intersection body for $k = n-3, n-2$. The authors do not know whether it is possible to extend the reverse inequalities to these values of k.

3 Main Results

The following theorem extends the reverse volume difference inequality (1.5) to the case of lower dimensional sections.

Theorem 3.1 *Let* $1 \leq k < n-3$, *and let* K *be an infinitely smooth origin-symmetric convex body in* $\mathbb{R}^n$, *with strictly positive curvature, that is not a* k-*intersection body. Then there exists an origin-symmetric convex body* L *in* $\mathbb{R}^n$ *such that* $L \subset K$ *and*

$$|K|^{\frac{n-k}{n}} - |L|^{\frac{n-k}{n}} < c_{n,k}^k \min_{H \in G_{n,n-k}} \left(|K \cap H| - |L \cap H| \right).$$

Proof Since K is an infinitely smooth convex body, $\| \cdot \|_K^{-k}$ is infinitely smooth. Thus $(\| \cdot \|_K^{-k})^\wedge$ is a continuous function on $\mathbb{R}^n \setminus \{0\}$. Since K is not a k-intersection body, by Proposition 2.4, there exists $\xi \in S^{n-1}$ such that $(\| \cdot \|_K^{-k})^\wedge(\xi) < 0$. By continuity of $(\| \cdot \|_K^{-k})^\wedge$, there exists a small neighborhood Ω of ξ where $(\| \cdot \|_K^{-k})^\wedge$ is still negative. We choose Ω smaller (if necessary) so that there exists a $(n-k)$-dimensional subspace H_0 such that $S^{n-1} \cap H_0^\perp \subset S^{n-1} \setminus \Omega$.

Let $\phi \in C^\infty(S^{n-1})$ be an even non-negative function on S^{n-1} with support in $\Omega \cup -\Omega$. Extend ϕ to an even homogeneous function $\phi \cdot r^{-k}$ of degree $-k$ on $\mathbb{R}^n$. The Fourier transform of the distribution $\phi \cdot r^{-k}$ is a homogeneous of degree $-n+k$ function $\psi \cdot r^{-n+k}$, where ψ is an infinitely smooth function on the sphere S^{n-1}.

Let ε be a positive number such that $|B_2^{n-k}| \cdot \|\theta\|_K^{-n+k} > \varepsilon > 0$ for all $\theta \in S^{n-1}$. Define a star body L by

$$\|\theta\|_L^{-n+k} = \|\theta\|_K^{-n+k} - \delta\psi(\theta) - \frac{\varepsilon}{|B_2^{n-k}|}, \quad \forall \theta \in S^{n-1}. \tag{3.1}$$

Select δ small enough so that for every θ

$$|\delta\psi(\theta)| < \min \left\{ \|\theta\|_K^{-n+k} - \frac{\varepsilon}{|B_2^{n-k}|}, \frac{\varepsilon}{|B_2^{n-k}|} \right\}.$$

This condition implies that L contains the origin and that $L \subset K$. Since L has strictly positive curvature, by the argument from [9, p.96], we can select ε, δ even smaller (if necessary) to ensure that the body L is convex.

Now we extend the functions in both sides of (3.1) to even homogeneous functions of degree $-n+k$ on $\mathbb{R}^n$ and apply the Fourier transform. We have that for every $\xi \in S^{n-1}$

$$(||\cdot||_L^{-n+k})^\wedge(\xi) = (||\cdot||_K^{-n+k})^\wedge(\xi) - (2\pi)^n\delta\phi(\xi) - \frac{\varepsilon}{|B_2^{n-k}|}(||\cdot||_2^{-n+k})^\wedge(\xi).$$

It follows from (2.5) that for every $\xi \in S^{n-1}$

$$(||\cdot||_L^{-n+k})^\wedge(\xi) = (||\cdot||_K^{-n+k})^\wedge(\xi) - (2\pi)^n\delta\phi(\xi) - \frac{(2\pi)^k(n-k)}{k\omega_k}\varepsilon. \qquad (3.2)$$

For every $H \in G_{n,n-k}$ we integrate (3.2) over $S^{n-1} \cap H^\perp$ and get

$$\int_{S^{n-1}\cap H^\perp}(||\cdot||_L^{-n+k})^\wedge(\xi)d\xi = \int_{S^{n-1}\cap H^\perp}(||\cdot||_K^{-n+k})^\wedge(\xi)d\xi - (2\pi)^n\delta\int_{S^{n-1}\cap H^\perp}\phi(\xi)d\xi$$
$$- (2\pi)^k(n-k)\varepsilon. \qquad (3.3)$$

Since $S^{n-1} \cap H_0^\perp \subset S^{n-1} \setminus \Omega$, we have $\phi(\xi) = 0$ for all $\xi \in H_0^\perp$. By (2.4), we get

$$\varepsilon = \min_{H\in G_{n,n-k}}(|K\cap H| - |L\cap H|). \qquad (3.4)$$

Multiplying both sides of (3.2) by $(||\cdot||_K^{-k})^\wedge$, integrating over the sphere S^{n-1} and using Parseval's formula on the sphere, we get

$$(2\pi)^n\int_{S^{n-1}}||\theta||_K^{-k}||\theta||_L^{-n+k}d\theta = (2\pi)^n n|K| - (2\pi)^n\delta\int_{S^{n-1}}\phi(\theta)(||\cdot||_K^{-k})^\wedge(\theta)d\theta$$
$$- \frac{(2\pi)^k(n-k)}{k\omega_k}\varepsilon\int_{S^{n-1}}(||\cdot||_K^{-k})^\wedge(\theta)d\theta.$$

Since ϕ is a non-negative function supported in $\Omega\cup-\Omega$, where $(||\cdot||_K^{-k})^\wedge$ is negative, the above inequality implies that

$$(2\pi)^n\int_{S^{n-1}}||\theta||_K^{-k}||\theta||_L^{-n+k}d\theta > (2\pi)^n n|K| - \frac{(2\pi)^k(n-k)}{k\omega_k}\varepsilon\int_{S^{n-1}}(||\cdot||_K^{-k})^\wedge(\theta)d\theta. \qquad (3.5)$$

By Hölder's inequality,

$$(2\pi)^n\int_{S^{n-1}}||\theta||_K^{-k}||\theta||_L^{-n+k}d\theta \le (2\pi)^n\left(\int_{S^{n-1}}||\theta||_K^{-n}d\theta\right)^{\frac{k}{n}}\left(\int_{S^{n-1}}||\theta||_L^{-n}d\theta\right)^{\frac{n-k}{n}}$$
$$= (2\pi)^n n|K|^{\frac{k}{n}}|L|^{\frac{n-k}{n}}. \qquad (3.6)$$

By (2.5), Parseval's formula on the sphere and Hölder's inequality,

$$\frac{(2\pi)^k(n-k)}{k\omega_k}\int_{S^{n-1}}(||\cdot||_K^{-k})^\wedge(\theta)d\theta = \frac{1}{|B_2^{n-k}|}\int_{S^{n-1}}(||\cdot||_K^{-k})^\wedge(\theta)(||\cdot||_2^{-n+k})^\wedge(\xi)d\theta$$

$$= \frac{(2\pi)^n}{|B_2^{n-k}|}\int_{S^{n-1}}||\theta||_K^{-k}d\theta$$

$$\leq \frac{(2\pi)^n}{|B_2^{n-k}|}|S^{n-1}|^{\frac{n-k}{n}}\left(\int_{S^{n-1}}||\theta||_K^{-n}d\sigma(\theta)\right)^{\frac{k}{n}}$$

$$= (2\pi)^n nc_{n,k}^k|K|^{\frac{k}{n}}. \tag{3.7}$$

Combining (3.5)–(3.7), we have

$$(2\pi)^n n|K| < (2\pi)^n n|K|^{\frac{k}{n}}|L|^{\frac{n-k}{n}} + (2\pi)^n nc_{n,k}^k|K|^{\frac{k}{n}}\varepsilon.$$

This, together with (3.4), implies the result.   □

The following result extends (1.4) to sections of lower dimensions.

Theorem 3.2 *Let* $1 \leq k < n$, *and let* L *be an infinitely smooth origin-symmetric convex body in* $\mathbb{R}^n$, *with strictly positive curvature, that is not a k-intersection body. Then there exists an origin-symmetric convex body* K *in* $\mathbb{R}^n$ *such that* $L \subset K$ *and*

$$|K|^{\frac{n-k}{n}} - |L|^{\frac{n-k}{n}} > \frac{C^k}{\left(\sqrt{n}M(\bar{L})\right)^k}\max_{H\in G_{n,n-k}}\left(|K\cap H| - |L\cap H|\right),$$

where $C > 0$ *is an absolute constant.*

Proof Since L is an infinitely smooth convex body, $||\cdot||_L^{-k}$ is infinitely smooth. Thus $(||\cdot||_L^{-k})^\wedge$ is a continuous function on $\mathbb{R}^n \setminus \{0\}$. Since L is not a k-intersection body, by Proposition 2.4, there exists $\xi \in S^{n-1}$ such that $(||\cdot||_L^{-k})^\wedge(\xi) < 0$. By continuity of $(||\cdot||_L^{-k})^\wedge$ there is a small neighborhood Ω of ξ where $(||\cdot||_L^{-k})^\wedge$ is negative. We choose Ω smaller (if necessary) so that there exists a $(n - k)$-dimensional subspace H_0 such that $S^{n-1} \cap H_0^\perp \subset S^{n-1} \setminus \Omega$.

Let $\phi \in C^\infty(S^{n-1})$ be an even non-negative function on S^{n-1} with support in $\Omega \cup -\Omega$. Extend ϕ to an even homogeneous function $\phi \cdot r^{-1}$ of degree $-k$ on $\mathbb{R}^n$. The Fourier transform of $\phi \cdot r^{-k}$ in the sense of distribution is $\psi \cdot r^{-n+k}$ where ψ is an infinitely smooth function on the sphere S^{n-1}.

Let ε be a positive number with $\varepsilon > 0$. Define a star body K by

$$||\theta||_K^{-n+k} = ||\theta||_L^{-n+k} - \delta\psi(\theta) + \frac{\varepsilon}{|B_2^{n-k}|}. \tag{3.8}$$

Choose δ small enough such that for every θ

$$|\delta\psi(\theta)| < \min\left\{||\theta||_L^{-n+k} + \frac{\varepsilon}{|B_2^{n-k}|}, \frac{\varepsilon}{|B_2^{n-k}|}\right\}.$$

This condition implies that K contains the origin and that $L \subset K$. Since L has strictly positive curvature, we also make ε, δ smaller (if necessary) to ensure that the body K is convex.

Now we extend the functions in (3.8) to even homogeneous functions of degree $-n+k$ on $\mathbb{R}^n$. Applying the Fourier transform we get for every $\xi \in S^{n-1}$

$$(||\cdot||_K^{-n+k})^\wedge(\xi) = (||\cdot||_L^{-n+k})^\wedge(\xi) - (2\pi)^n\delta\phi(\xi) + \frac{\varepsilon}{|B_2^{n-k}|}(||\cdot||_2^{-n+k})^\wedge(\xi).$$

Hence, it follows from (2.5) that for every $\xi \in S^{n-1}$

$$(||\cdot||_K^{-n+k})^\wedge(\xi) = (||\cdot||_L^{-n+k})^\wedge(\xi) - (2\pi)^n\delta\phi(\xi) + \frac{(2\pi)^k(n-k)}{k\omega_k}\varepsilon. \qquad (3.9)$$

For every $H \in G_{n,n-k}$ we integrate (3.9) over $S^{n-1} \cap H^\perp$ and get

$$\int_{S^{n-1}\cap H^\perp}(||\cdot||_K^{-n+k})^\wedge(\xi)d\xi = \int_{S^{n-1}\cap H^\perp}(||\cdot||_L^{-n+k})^\wedge(\xi)d\xi - (2\pi)^n\delta\int_{S^{n-1}\cap H^\perp}\phi(\xi)d\xi$$

$$+ (2\pi)^k(n-k)\varepsilon. \qquad (3.10)$$

Since $S^{n-1} \cap H_0^\perp \subset S^{n-1} \setminus \Omega$, we have $\phi(\xi) = 0$ for $\xi \in H_0^\perp$. By (2.4), we have that

$$\varepsilon = \max_{H\in G_{n,n-k}}(|K \cap H| - |L \cap H|). \qquad (3.11)$$

Multiplying both sides of (3.9) by $(||\cdot||_L^{-k})^\wedge$, integrating over the sphere S^{n-1} and using Parseval's formula on the sphere, we get

$$(2\pi)^n\int_{S^{n-1}}||\theta||_L^{-k}||\theta||_K^{-n+k}d\theta = (2\pi)^n n|L| - (2\pi)^n\delta\int_{S^{n-1}}\phi(\theta)(||\cdot||_L^{-k})^\wedge(\theta)d\theta$$

$$+ \frac{(2\pi)^k(n-k)}{k\omega_k}\varepsilon\int_{S^{n-1}}(||\cdot||_L^{-k})^\wedge(\theta)d\theta.$$

Since ϕ is a non-negative function with support in $\Omega \cup -\Omega$, where $(||\cdot||_L^{-k})^\wedge$ is negative, the above inequality implies that

$$(2\pi)^n\int_{S^{n-1}}||\theta||_L^{-k}||\theta||_K^{-n+k}d\theta > (2\pi)^n n|L| + \frac{(2\pi)^k(n-k)}{k\omega_k}\varepsilon\int_{S^{n-1}}(||\cdot||_L^{-k})^\wedge(\theta)d\theta. \qquad (3.12)$$

By Hölder's inequality,

$$(2\pi)^n \int_{S^{n-1}} ||\theta||_L^{-k} ||\theta||_K^{-n+k} d\theta \leq (2\pi)^n n |L|^{\frac{k}{n}} |K|^{\frac{n-k}{n}}. \tag{3.13}$$

By (2.5), Parseval's formula on the sphere and Jensen's inequality,

$$\frac{(2\pi)^k(n-k)}{k\omega_k} \int_{S^{n-1}} (||\cdot||_L^{-k})^\wedge(\theta) d\theta = \frac{1}{|B_2^{n-k}|} \int_{S^{n-1}} (||\cdot||_L^{-k})^\wedge(\theta)(||\cdot||_2^{-n+k})^\wedge(\xi) d\theta$$

$$= \frac{(2\pi)^n |S^{n-1}|}{|B_2^{n-k}|} \int_{S^{n-1}} ||\theta||_L^{-k} d\sigma(\theta)$$

$$\geq \frac{(2\pi)^n |S^{n-1}|}{|B_2^{n-k}|} \left(\frac{1}{\int_{S^{n-1}} ||\theta||_L d\sigma(\theta)} \right)^k$$

$$= \frac{(2\pi)^n |S^{n-1}|}{|B_2^{n-k}|} \frac{|L|^{\frac{k}{n}}}{M(\bar{L})^k}$$

$$\geq (2\pi)^n n C^k \frac{|L|^{\frac{k}{n}}}{\left(\sqrt{n} M(\bar{L})\right)^k}. \tag{3.14}$$

From standard estimates for the Γ-function and (3.12)–(3.14), we get

$$(2\pi)^n n |L|^{\frac{k}{n}} |K|^{\frac{n-k}{n}} > (2\pi)^n n |L| + (2\pi)^n n C^k \frac{|L|^{\frac{k}{n}}}{\left(\sqrt{n} M(\bar{L})\right)^k} \varepsilon.$$

This, together with (3.11), implies the required result. □

Acknowledgements The first named author was supported in part by the US National Science Foundation grant DMS-1700036. The second named author was partially supported by Fundamental Research Funds for the Central Universities (No. XDJK2016D026) and China Scholarship Council.

References

1. S. Artstein, A. Giannopoulos, V. Milman, *Asymptotic Geometric Analysis*. Mathematical Surveys and Monographs, vol. 202 (American Mathematical Society, Providence, RI, 2015)
2. I.M. Gelfand, G.E. Shilov, *Generalized Functions*. Properties and Operations, vol. 1 (Academic Press, New York, 1964)
3. A. Giannopoulos, A. Koldobsky, Variants of the Busemann-Petty problem and of the Shephard problem. Int. Math. Res. Not. **2017**(3), 921–943 (2017)
4. A. Giannopoulos, A. Koldobsky, Volume difference inequalities. Trans. Am. Math. Soc. (to appear). https://doi.org/10.1090/tran/7173
5. P. Goodey, W. Weil, Intersection bodies and ellipsoids. Mathematika **42**, 295–304 (1995)

6. E. Grinberg, G. Zhang, Convolutions, transforms and convex bodies. Proc. Lond. Math. Soc. (3) **78**, 77–115 (1999)
7. A. Koldobsky, An application of the Fourier transform to sections of star bodies. Isr. J. Math. **106**, 157–164 (1998)
8. A. Koldobsky, A functional analytic approach to intersection bodies. Geom. Funct. Anal. **10**, 1507–1526 (2000)
9. A. Koldobsky, *Fourier Analysis in Convex Geometry*. Mathematical Surveys and Monographs, vol. 116, (American Mathematical Society, Providence, RI, 2005)
10. A. Koldobsky, Stability in the Busemann-Petty and Shephard problems. Adv. Math. **228**, 2145–2161 (2011)
11. A. Koldobsky, Slicing inequalities for measures of convex bodies. Adv. Math. **283**, 473–488 (2015)
12. A. Koldobsky, M. Lifshits, Average volume of sections of star bodies, in *Geometric Aspects of Functional Analysis*, ed. by V. Milman, G. Schechtman. Lecture Notes in Mathematics, vol. 1745 (Springer, Berlin, 2000), pp. 119–146
13. A. Koldobsky, D. Ma, Stability and slicing inequalities for intersection bodies. Geom. Dedicata **162**, 325–335 (2013)
14. E. Lutwak, Intersection bodies and dual mixed volumes. Adv. Math. **71**(2), 232–261 (1988)
15. G. Zhang, Sections of convex bodies. Am. J. Math. **118**, 319–340 (1996)

Around the Simplex Mean Width Conjecture

Alexander E. Litvak

Abstract In this note we discuss an old conjecture in Convex Geometry asserting that the regular simplex has the largest mean width among all simplices inscribed into the Euclidean ball and its relation to Information Theory. Equivalently, in the language of Gaussian processes, the conjecture states that the expectation of the maximum of $n + 1$ standard Gaussian variables is maximal when the expectations of all pairwise products are $-1/n$, that is, when the Gaussian variables form a regular simplex in L_2. We mention other conjectures as well, in particular on the expectation of the smallest (in absolute value) order statistic of a sequence of standard Gaussian variables (not necessarily independent), where we expect the same answer.

1 Introduction

By B_2^n and S^{n-1} we denote the standard unit Euclidean ball and the unit Euclidean sphere in $\mathbb{R}^n$. Then $|\cdot|$ and $\langle \cdot, \cdot \rangle$ denote the corresponding Euclidean norm and inner product. By $\{e_i\}_{i=1}^n$ we denote the canonical basis of $\mathbb{R}^n$ and by Δ_n we denote the regular simplex inscribed into S^{n-1}. Given a convex body K in $\mathbb{R}^n$, its support function and mean width are defined by

$$ h_K(u) = \max_{x \in K} \langle u, x \rangle \qquad \text{and} \qquad w(K) = 2 \int_{S^{n-1}} h_K(u) \, d\mu(u), $$

where μ is the normalized Lebesgue measure on S^{n-1}.

In this note we discuss the following long-standing open conjecture and some related results and conjectures.

A. E. Litvak (✉)
Department of Mathematical and Statistical Sciences, University of Alberta, Edmonton, AB, Canada
e-mail: aelitvak@gmail.com

© Springer International Publishing AG 2018
G. Bianchi et al. (eds.), *Analytic Aspects of Convexity*, Springer INdAM Series 25,
https://doi.org/10.1007/978-3-319-71834-7_5

Conjecture 1.1 Among all simplices inscribed into B_2^n the regular simplex Δ_n has the maximal mean width. Moreover, Δ_n is the unique simplex maximizing mean width.

This conjecture was briefly discussed in a survey of Gritzmann and Klee ([21], Section 9.10.2) and was mentioned several times by Klee in his talks. It was also mentioned in Böröczky's book [10] and in recent works by Hug and Schneider [22] and by Böröczky and Schneider [11]. We refer to [9] for related results on the mean width of a simplex and to [37] for the general theory of convex bodies and Brunn–Minkowski Theory. In the Information Theory community it was a general belief that the conjecture is known to be true (see e.g. [2, 3, 14, 41, 46]). We discuss the importance of Conjecture 1.1 to Information Theory in Sect. 2.

We would like to emphasize an important difference in the setting of this problem with a standard setting of problems in Asymptotic Geometric Analysis, where we usually identify bodies which can be obtained from each other by an affine transformation (or, in the centrally symmetric case, by a linear transformation), working thus with equivalence classes of bodies. In many problems we usually fix *a position* of a body, where by a position of a body we understand a certain affine (linear in the centrally symmetric case) image of it. In this context Conjecture 1.1 has to be compared with the following results by Barthe [8] (who developed the approach originated by Ball in [5, 6] to describe bodies with the maximal volume ratio and surface area) and by Schmuckenschläger [36] respectively, investigating maximizers and minimizers of mean width of bodies in John's and Löwner's positions.

Theorem 1.2 *Among all convex bodies in the John position, that is, bodies, whose maximal volume ellipsoid is the unit Euclidean ball B_2^n, the regular simplex $n\Delta_n$ has the largest mean width.*

Theorem 1.3 *Among all convex bodies in the Löwner position, that is, bodies, whose minimal volume ellipsoid is the unit Euclidean ball B_2^n, the regular simplex Δ_n has the smallest mean width.*

Corresponding statements for the class of centrally symmetric bodies were proved by Schechtman and Schmuckenschläger [35]. The maximizer of the mean width among all centrally symmetric bodies in the John position is the cube, while the minimizer of the mean width among all centrally symmetric bodies in the Löwner position is the cross-polytope (octahedron).

We now reformulate the conjecture in the language of Gaussian processes. By $g_1, g_2, g_3, \ldots$ we always denote i.i.d. standard Gaussian random variables ($g_i \sim \mathcal{N}(0, 1)$). $G = (g_1, \ldots, g_n)$ denotes the standard Gaussian random vector in $\mathbb{R}^n$. The integration in polar coordinates leads to

$$w(K) = c_n \, \mathbb{E} \, h_K(G), \tag{1}$$

where c_n is a constant depending only on n (in fact, $c_n \geq \frac{2}{\sqrt{n}}$ and $c_n\sqrt{n} \to 2$).

When $K = \text{conv}\{x_1, \ldots, x_{n+1}\} \subset \mathbb{R}^n$, $|x_i| = 1$ for $i \leq n+1$, we have

$$\mathbb{E}\, h_K(G) = \mathbb{E} \max_{i \leq n+1} \langle G, x_i \rangle.$$

Denote $\xi_i = \langle G, x_i \rangle$, $i \leq n+1$. Then $\xi_i \sim \mathcal{N}(0, 1)$ and

$$\sigma_{ij} = \sigma_{ij}(K) := \mathbb{E}\xi_i \xi_j = \langle x_i, x_j \rangle.$$

Recall, that the vertices $v_1, \ldots, v_{n+1}$ of Δ_n satisfy

$$|v_i| = 1 \qquad \text{and} \qquad \langle v_i, v_j \rangle = -\frac{1}{n} \quad \text{for all} \quad i \neq j.$$

The $(n+1) \times (n+1)$ covariance matrix corresponding to the regular simplex, that is, $\sigma = \{\sigma_{ij}\}_{ij}$ with $\sigma_{ii} = 1$ and $\sigma_{ij} = -\frac{1}{n}$ for $i \neq j$, we denote by $\sigma(\Delta_n)$. Thus Conjecture 1.1 is equivalent to

Conjecture 1.4 Among all Gaussian random vectors $(\xi_1, \ldots, \xi_{n+1})$ with $\xi_i \sim \mathcal{N}(0, 1)$ for all $i \leq n+1$, the expectation

$$\mathbb{E} \max_{i \leq n+1} \xi_i$$

is maximal when the covariance matrix $\sigma = \sigma(\Delta_n)$. The solution is unique.

We would like to emphasize that if we add the absolute values to ξ_i's, that is, if we want to maximize $\mathbb{E} \max_{i \leq n+1} |\xi_i|$, then the answer is known—the maximum of such expectation attains when ξ_i's are independent as was proved by Šidák [39] and Gluskin [18]. Geometrically it says the following.

Theorem 1.5 *Among all linear images of the cross-polytope contained in B_2^n the cross-polytope itself has the maximal mean width.*

Indeed, denote the cross-polytope by B_1^n and let T be a linear operator such that $TB_1^n \subset B_2^n$. Denote $x_i = Te_i$ for $i \geq 1$. Without loss of generality, $|x_i| = 1$. Let $\xi_i = \langle G, x_i \rangle$, $i \geq 1$. Then $\xi_i \sim \mathcal{N}(0, 1)$. Therefore, by (1) and by the Šidák theorem,

$$w(TB_1^n) = c_n \mathbb{E} \max_{i \leq n} |\xi_i| \leq c_n \mathbb{E} \max_{i \leq n} |g_i| = w(B_1^n).$$

The other counterpart of this theorem follows from Proposition 4 in [31] in a similar way, namely we have

Theorem 1.6 *Among all linear images of the cross-polytope containing $\frac{1}{\sqrt{n}} B_2^n$ the cross-polytope itself has the minimal mean width.*

Balitskiy, Karasev, and Tsigler conjectured that for every $r > 0$ the Gaussian measure of a simplex S containing rB_2^n is minimized when S is regular (see Conjecture 3.3 in [4]). In the same way as Šidák's theorem and Proposition 4 in [31] imply Theorems 1.5 and 1.6, their conjecture would immediately imply Conjecture 1.1. Moreover, they showed that their conjecture implies Conjecture 2.4 formulated below.

This note is organized as follows. In Sect. 2, we discuss the *Simplex Code Conjecture* or the *Weak Simplex Conjecture*, which makes links between Conjecture 1.1 and Information theory. In particular, we mention a problem, Problem 2.2, on the behaviour of the maximum of a certain Gaussian process which is (if solution is a regular simplex) stronger than Conjecture 1.4 and which is needed to solve the corresponding problem in transmitting signals. In Sect. 3, we describe two other stronger conjectures related to the Steiner formula and to the intrinsic volumes. Then, in Sect. 4, we provide some asymptotic estimates, in particular we show that the mean width of a half-dimensional (flat) cross-polytope is surprisingly very close to the mean width of a regular simplex. We also show that the regular simplex is a solution in the asymptotic sense. Finally, in Sect. 5, we formulate another conjecture on the minimum of a Gaussian process, where, as we believe, the solution is also the regular simplex. It seems that Conjectures 3.1, 3.2, 5.1 have not appeared in the literature yet.

2 Simplex Code Conjecture

In this section we discuss the relation of Conjecture 1.1 to Information Theory. We first describe a problem in transmitting signals which goes back to works of Kotel'nikov [25], MacColl [33], and Shannon [38] published at the end of 40-s.

For positive integers n and N let $x_1, \ldots, x_N \in S^{n-1}$ be signal vectors to be transmitted. Let $Y = \lambda x_i + G$ be the observed (received) vector when x_i was transmitted. Here, G is the standard Gaussian random vector (corresponding to the white noise) and $\lambda > 0$ is the (fixed) signal-to-noise ratio. The problem is:

Problem 2.1 Observed Y to reconstruct x_j which has been transmitted.

To solve the problem one creates the *matched filter*—the vectors $y_i = \langle Y, x_i \rangle$, $i \leq N$, and decides that x_j was transmitted if

$$y_j = \max_{i \leq N} y_i.$$

We want to maximize the probability of the right decision, that is, to maximize the function

$$\psi_\lambda \left(\{x_i\}_{i=1}^N \right) = \frac{1}{N} \sum_{j=1}^N \mathbb{P} \left(y_j = \max_{i \leq N} \langle Y, x_i \rangle \ \Big| \ Y = \lambda x_j + G \right).$$

In his works [1, 3] Balakrishnan essentially developed the theory (see also Chapter 14 of Weber's [46] book for self-contained presentation). In [1] Balakrishnan proved that the latter problem of maximizing ψ_λ is equivalent to the following problem.

Problem 2.2 Given a Gaussian random vector $(\xi_1, \ldots, \xi_N)$ with $\xi_i \sim \mathcal{N}(0, 1)$ for all $i \leq N$ and a covariance matrix σ of rank n set

$$\phi_\lambda(\sigma) := \mathbb{E} \exp\left(\lambda \max_{i \leq N} \xi_i\right).$$

Maximize $\phi_\lambda(\sigma)$ over all choices of covariance matrices σ.

Differentiating with respect to λ, one gets the following [1].

Lemma 2.3 *If there exists a solution of Problem 2.2 which is independent of λ in an interval $(0, \lambda_0)$ for some $\lambda_0 > 0$, then this solution maximizes mean width of the convex hull of x_i's.*

We turn now to the case of the simplex, that is, we fix $N = n + 1$. The following is the Simplex Code Conjecture or the Weak Simplex Conjecture.

Conjecture 2.4 Let $N = n + 1$. For every (fixed) $\lambda > 0$ the function $\phi_\lambda(\cdot)$ is maximal for the regular simplex, that is, when the covariance matrix of ξ_i's is $\sigma = \sigma(\Delta_n)$.

The following two theorems were proved in [1] and [3] respectively.

Theorem 2.5 *Let $N = n + 1$. For every (fixed) $\lambda > 0$, the regular simplex is a local maximum in Problem 2.2. Furthermore, if there exists a solution of Problem 2.2 which is independent of λ on some interval (λ_1, λ_2), then this solution is given by the regular simplex.*

Theorem 2.6 *Given $\lambda > 0$ let σ_λ denote a point (or one of points) of maximum of $\phi_\lambda(\cdot)$. Then*

$$\lim_{\lambda \to \infty} \sigma_\lambda = \sigma(\Delta_n).$$

In [3] a similar statement about the limit at 0 is claimed but its proof essentially uses Conjecture 1.1.

In 1966 Landau and Slepian [26] published a proof of Conjecture 2.4 for arbitrary $n \geq 3$. The proof was based on the geometric technique developed by Fejes-Tóth ([16], pp. 137–138). However, as was noticed by Farber [15] and Tanner [44], the 3-dimensional proofs of corresponding geometric results in [16] do not extend to higher dimensions (see also [4] for more insight and related conjectures). Therefore the proof of Conjecture 2.4 in [26] holds only for $n = 3$. We refer to [45] for two more related conjectures, which are stronger than Conjecture 2.4. We would also like to mention that Cover [13] noticed that Conjecture 2.4 can be reduced to the following geometric problem.

Conjecture 2.7 Let $A > 0$ and $B \subset S^{n-1}$ be a spherical cap with the center at $x \in S^{n-1}$. Let S be a spherical simplex of the area A, where by a spherical simplex we understand the intersection of the sphere S^{n-1} with n half-spaces. Then a regular spherical simplex with the center at x maximizes the area of the intersection $B \cap S$.

Finally we would like to mention the Strong Simplex Conjecture, which asserts the same as the Weak Simplex Conjecture, but instead of constraints $x_i \in S^{n-1}$, $i \leq n + 1$, in the initial problem one uses the constraint

$$\sum_{i=1}^{n+1} |x_i|^2 = n + 1,$$

i.e., instead of choosing x_i's on the sphere we fix the sum of their squared lengths. The Strong Simplex Conjecture was disproved by Steiner [41], who used essentially a one-dimensional example. For other counter-examples see [27, 43].

3 Two More Geometric Conjectures

In this section we discuss connections of Conjecture 1.1 to the Steiner formula and to the intrinsic volumes and provide two more geometric conjectures which are stronger than Conjecture 1.1. Both conjectures were communicated to us after the first draft of this note was written. We refer to [37] and references therein for the general theory of convex bodies, Brunn–Minkowski Theory, in particular for information about and relations between the mean width functional, quermassintegrals, intrinsic volumes, etc. (see also [42] for relations between intrinsic volumes and Gaussian processes).

Let $t > 0$ and K be a convex body in $\mathbb{R}^n$. Consider the Minkowski sum

$$K_t := K + tB_2^n = \{x + ty \mid x \in K, \, y \in B_2^n\} = \{z \mid \mathrm{dist}(z, K) \leq t\},$$

where dist denotes the Euclidean distance. The Steiner formula says that the (n-dimensional) volume $|K_t|$ of K_t is polynomial in t. It can be written as

$$|K_t| = \sum_{j=0}^{n} \kappa_{n-j} V_j(K) t^{n-j},$$

where $\kappa_0 = 1$, $\kappa_i = |B_2^i| = \pi^{i/2} / \Gamma(1 + i/2)$ for $i \geq 1$, and $V_i(K)$, $i \leq n$, are coefficients which depend only on K. These coefficients are called intrinsic volumes of K. Analyzing the Steiner formula it is easy to see that $V_0(K) = 1$, $V_n(K) = |K|$, and $2V_{n-1}(K)$ is the surface area of K. Moreover, it is known that $V_1(K) = n\kappa_n/(2\kappa_{n-1}) w(K)$ and $V_1 = \sqrt{2\pi}\mathbb{E} \max_{x \in K} \sup \langle G, x \rangle$ (see e.g. Proposition 2.4.14 in [42], cf. (1)).

R. van Handel (private communications) suggested the following natural extension of Conjecture 1.1.

Conjecture 3.1 For every (fixed) $t > 0$, among all simplices $S \subset B_2^n$ the regular simplex Δ_n maximizes the volume $|S_t|$. Moreover, Δ_n is the unique simplex maximizing this volume.

Conjecture 3.1 is stronger than Conjecture 1.1. Indeed, using the Steiner formula and that $V_0(K) = 1$ for every K, we have

$$|(\Delta_n)_t| - |S_t| = \sum_{j=0}^{n-1} \kappa_j \left(V_{n-j}(\Delta_n) - V_{n-j}(S) \right) t^j,$$

therefore, sending t to infinity, we observe $V_{n-1}(\Delta_n) \geq V_{n-1}(S)$.

Furthermore, Z. Kabluchko and D. Zaporozhets (private communications) suggested even stronger conjecture.

Conjecture 3.2 Let $S \subset B_2^n$ be a simplex. Then for every $1 \leq i \leq n$ one has $V_i(S) \leq V_i(\Delta_n)$. Moreover, if S is not regular, then the inequality is strict for every i.

We would like to note that some cases in Conjecture 3.2 are known. Indeed, the case $i = n$ corresponds to the volume. It follows from the John theorem ([7, 23], see also [45] for a geometric proof). The case $i = n - 1$, corresponding to the surface area, was proved by Tanner in [45].

4 Asymptotic Results and Comparison to Cross-Polytope

In this section we discuss asymptotic behaviour and compare mean width of the regular simplex with the mean width of corresponding half-dimensional cross-polytope, showing that surprisingly they are very close to each other. All results of this section with complete proofs can be found in [24].

As before, let g_i's denote i.i.d. standard Gaussian random variables. Let $(\eta_1, \ldots, \eta_{n+1})$ denote a Gaussian random vector with the covariance matrix $\sigma(\Delta_n)$.

We believe that the following observation has been known for many years. Balakrishnan [1] tributes it to C.R. Chan (in the context of the function ψ_λ defined in the previous section). It can be obtained by direct calculations, since Δ_n can be realized as the (properly normalized) convex hull of the canonical basis $\{e_i\}_{i=1}^{n+1}$ in $\mathbb{R}^{n+1}$.

Claim 4.1

$$\mathbb{E} \max_{i \leq n+1} \eta_i = \sqrt{\frac{n+1}{n}} \, \mathbb{E} \max_{i \leq n+1} g_i = \left(1 + \frac{1 + o(1)}{2n} \right) \mathbb{E} \max_{i \leq n+1} g_i.$$

The next statement claims that the regular simplex is the best asymptotically. Its proof is based on standard estimates for Gaussian processes.

Lemma 4.2 *If $S_n \subset B_2^n$ is a simplex with the maximal mean width then*

$$w(\Delta_n) \leq w(S_n) \leq \left(1 + \frac{C \ln \ln n}{\ln n}\right) w(\Delta_n),$$

where C is a positive absolute constant.

Note here that in [17] the mean width of the regular simplex was calculated as

$$w(\Delta_n) = 2\sqrt{\frac{\ln n}{n}} \left(1 - (1 + o(1)) \frac{\ln \ln n}{\ln n}\right).$$

We turn now to the comparison with the cross-polytope (octahedron). Recall that we consider convex hulls of $n + 1$ points on the Euclidean sphere. Assume that $n = 2m - 1$ and consider the m-dimensional cross-polytope $B_1^m = \text{conv}\{\pm e_i\}_{i=1}^m$ in $\mathbb{R}^n$. Clearly, B_1^m is a (degenerated) simplex in $\mathbb{R}^n$. Surprisingly, the mean width of B_1^m is very close to the mean width of Δ_n as the next theorem shows (recall that the mean width can be computed via corresponding expectations of Gaussian processes and that Claim 4.1 relates dependent Gaussian random variables corresponding to the regular simplex with independent Gaussian random variables). The left hand side inequality in Theorem 4.3 is immediate by Slepian's Lemma ([40], see also [28, 30]).

Theorem 4.3 *Let $n = 2m - 1$. Then*

$$\mathbb{E} \max_{i \leq 2m} g_i \leq \mathbb{E} \max_{i \leq m} |g_i| = \left(1 + \frac{1 + o(1)}{4n \ln n}\right) \mathbb{E} \max_{i \leq n+1} g_i.$$

In particular,

$$w(B_1^m) = \left(1 - \frac{1 + o(1)}{2n}\right) w(\Delta_n).$$

Moreover, using the Chatterjee technique ([12]), a path σ_t, $t \in [0, 1]$, in covariance matrices of Gaussian vectors can be constructed so that σ_0 corresponds to the vector

$$\sqrt{\frac{n+1}{n}} (g_1, \ldots, g_{n+1})$$

(that is, to the regular simplex) and σ_1 corresponds to the vector

$$(g_1, -g_1, g_2, -g_2, \ldots, g_m, -g_m)$$

(that is, to the m-dimensional cross-polytope) and such that the expectation of maximum is not-decreasing along this path. This gives the following estimate, which is slightly weaker for large n, but better for small n.

Theorem 4.4 *Let* $n = 2m - 1$.

$$\mathbb{E} \max_{i \leq 2m} g_i \leq \mathbb{E} \max_{i \leq m} |g_i| \leq \sqrt{\frac{n+1}{n}} \, \mathbb{E} \max_{i \leq 2m} g_i.$$

In particular,

$$w(B_1^m) \leq w(\Delta_n) \leq \sqrt{\frac{n+1}{n}} \, w(B_1^m).$$

5 A Conjecture on the Smallest Order Statistic

In this section we formulate a conjecture on Gaussian processes, which, to the best of our knowledge, appears for the first time. Although it is not directly related to the mean width of convex bodies, we have decided to mention it here, since it also deals with an extreme order statistic of coordinates of the standard Gaussian vector (cf. Conjecture 1.4) and since we believe that it has the same solution as Conjecture 1.1.

Conjecture 5.1 Let $n \geq 2$ and $p > 0$. Among all Gaussian random vectors $(\xi_1, \ldots, \xi_{n+1})$ with $\xi_i \sim \mathcal{N}(0, 1)$ for all $i \leq n + 1$, the expectation

$$\mathbb{E} \min_{i \leq n+1} |\xi_i|^p$$

is minimal when the covariance matrix $\sigma = \sigma(\Delta_n)$. The solution is unique.

The main motivation for this question comes from the Mallat-Zeitouni problem which is still open in full generality ([34], see also [32] for discussions, history, references, and a partial solution). In the original notes, published on Zeitouni's webpage in 2000, Mallat and Zeitouni suggested a way to solve it. Their method would have worked if a more general result in the spirit of Conjecture 5.1 had held with $p = 2$. Moreover, in [19, 20] the authors were able to prove that for every sequence of real numbers $\{a_i\}_{i=1}^{n+1}$ and every $p > 0$,

$$\mathbb{E} \min_{i \leq n+1} |a_i g_i|^p \leq \Gamma(2 + p) \, \mathbb{E} \min_{i \leq n+1} |a_i \xi_i|^p,$$

where $\Gamma(\cdot)$ is the Gamma-function. This result, together with the Šidák theorem also supported the intuition that the independent case gives the minimum (see also [29]). However later, van Handel cheked that for $n = 2$ the arrangement corresponding for Δ_2 is better than three independent variables [34].

Acknowledgements The author is grateful to A. Akopyan, Z. Kabluchko, R. Karasev, J. Prochno, R. Schneider, R. van Handel, D. Zaporozhets for valuable remarks on the first draft of this note and to Z. Kabluchko, R. van Handel, and D. Zaporozhets for bringing conjectures described in Sect. 3 to his attention.

References

1. A.V. Balakrishnan, A contribution to the sphere-packing problem of communication theory. J. Math. Anal. Appl. **3**, 485–506 (1961)
2. A.V. Balakrishnan, Research problem no. 9: geometry. Bull. Am. Math. Soc. **69**, 737–738 (1963)
3. A.V. Balakrishnan, Signal selection for space communication channels, in *Advances in Communication Systems*, ed. by A.V. Balakrishnan (Academic Press, New York, 1965), pp. 1–31
4. A. Balitskiy, R. Karasev, A. Tsigler, Optimality of codes with respect to error probability in Gaussian noise (2017). https://arxiv.org/abs/1701.07986
5. K. Ball, Volumes of sections of cubes and related problems, in *Geometric Aspects of Functional Analysis (1987–1988)*. Lecture Notes in Mathematics, vol. 1376 (Springer, Berlin, 1989), pp. 251–260
6. K. Ball, Volume ratios and a reverse isoperimetric inequality. J. Lond. Math. Soc. (2) **44**(2), 351–359 (1991)
7. K. Ball, Ellipsoids of maximal volume in convex bodies. Geom. Dedicata **41**, 241–250 (1992)
8. F. Barthe, An extremal property of the mean width of the simplex. Math. Ann. **310**, 685–693 (1998)
9. K. Böröczky Jr., About the mean width of simplices. Period. Polytech. Mech. Eng. **36**, 291–297 (1992)
10. K. Böröczky Jr., *Finite Packing and Covering*. Cambridge Tracts in Mathematics, vol. 154 (Cambridge University Press, Cambridge, 2004)
11. K. Böröczky Jr., R. Schneider, Circumscribed simplices of minimal mean width. Beitr. Algebra Geom. **48**, 217–224 (2007)
12. S. Chatterjee, An error bound in the Sudakov–Fernique inequality (2005). http://arxiv.org/abs/math/0510424
13. T.M. Cover, B. Gopinath (eds.), *Open Problems in Communication and Computation* (Springer, New York, 1987)
14. B. Dunbridge, Asymmetric signal design for the coherent Gaussian channel. IEEE Trans. Inf. Theory **IT-13**, 422–431 (1967)
15. S.M. Farber, On the signal selection problem for phase coherent and incoherent communication channels. Tech. Report No. 4, Communications Theory Lab., Dept. of Electrical Engineering, California Institute of Technology (1968)
16. L. Fejes Tóth, *Lagerungen in der Ebene, auf der Kugel und im Raum*. Die Grundlehren der Mathematischen Wissenschaften in Einzeldarstellungen mit besonderer Berücksichtigung der Anwendungsgebiete, Band LXV (Springer, Berlin, Göttingen, Heidelberg, 1953)
17. S.R. Finch, Mean width of a regular simplex (2011). http://arxiv.org/abs/1111.4976
18. E.D. Gluskin, Extremal properties of orthogonal parallelepipeds and their applications to the geometry of Banach spaces. Math. USSR Sb. **64**, 85–96 (1989)
19. Y. Gordon, A.E. Litvak, C. Schütt, E. Werner, Minima of sequences of Gaussian random variables. CR Acad. Sci. Paris Sér. I Math. **340**, 445–448 (2005)
20. Y. Gordon, A.E. Litvak, C. Schütt, E. Werner, On the minimum of several random variables. Proc. Am. Math. Soc. **134**, 3665–3675 (2006)

21. P. Gritzmann, V. Klee, On the complexity of some basic problems in computational convexity II: volume and mixed volumes, in *Polytopes: Abstract, Convex and Computational. Proceedings of the NATO Advanced Study Institute*, ed. by T. Bisztriczky et al. Nato Science Series C: Mathematical and Physical Sciences, vol. 440 (Kluwer Academic, Boston, 1994), pp. 373–466.

22. D. Hug, R. Schneider, Large typical cells in Poisson-Delaunay mosaics. Rev. Roum. Math. Pures Appl. **50**, 657–670 (2005)

23. F. John, Extremum problems with inequalities as subsidiary conditions, in *Studies and Essays Presented to R. Courant on his 60th Birthday, January 8, 1948* (Interscience Publishers, Inc., NY, 1948), pp. 187–204

24. Z. Kabluchko, A.E. Litvak, D. Zaporozhets, On the expected maximum of correlated Gaussian variables. Zapiski Nauchnykh Seminarov POMI **442**, 75–96 (2015); English translation in J. Math. Sci. **225**, 770–787 (2017)

25. V.A. Kotel'nikov, *Thesis*, Molotov Energy Institute, Moscow, 1947, translated by R.A. Silverman, *The Theory of Optimum Noise Immunity* (McGraw-Hill, New York, 1959)

26. H.J. Landau, D. Slepian, On the optimality of the regular simplex code. Bell Syst. Tech. J. **45**, 1247–1272 (1966)

27. D.E. Lazich, C. Senger, M. Bossert, A corrected disproof of the strong simplex conjecture, in *SCC 2013; 9th International ITG Conference on Systems, Communication and Coding* (2013)

28. M. Ledoux, M. Talagrand, *Probability in Banach spaces*. Isoperimetry and Processes (Springer, Berlin, 1991)

29. W. Li, *L1: Introduction, Overview and Applications*. CBMS Lectures at UAH, June 4–8, 2012. http://jamesyli.com/wenboli_backup/papers/CBMS-10Lectures.pdf

30. M.A. Lifshits, *Gaussian Random Functions*. Mathematics and Its Applications, vol. 322 (Kluwer Academic, Dordrecht, 1995)

31. A.E. Litvak, P. Mankiewicz, N. Tomczak-Jaegermann, Randomized isomorphic Dvoretzky theorem. CR Acad. Sci. Paris Ser. 1 Math. **335**, 345–350 (2002)

32. A.E. Litvak, K. Tikhomirov, Order statistics of vectors with dependent coordinates, and the Karhunen–Loeve basis. Ann. Appl. Probab. (to appear)

33. L.A. MacColl, *Signalling in the Presence of Thermal Noise, I, II, and III*. Bell Laboratories internal memoranda issued May 27, June 30, and September 13, 1948

34. S. Mallat, O. Zeitouni, A conjecture concerning optimality of the Karhunen-Loeve basis in nonlinear reconstruction (2011). arXiv: 1109.0489

35. G. Schechtman, M. Schmuckenschläger, A concentration inequality for harmonic measures on the sphere, in *Geometric Aspects of Functional Analysis (Israel, 1992–1994)*. Operator Theory: Advances and Applications, vol. 77 (Birkhäuser, Basel, 1995), pp. 255–273

36. M. Schmuckenschläger, An extremal property of the regular simplex, in *Convex Geometric Analysis (Berkeley, CA, 1996)*. Mathematical Sciences Research Institute Publications, vol. 34 (Cambridge University Press, Cambridge, 1999), pp. 199–202

37. R. Schneider, *Convex Bodies: the Brunn-Minkowski Theory*, second expanded edition. Encyclopedia of Mathematics and Its Applications, vol. 151 (Cambridge University Press, Cambridge, 2014)

38. C.E. Shannon, Communication in the presence of noise. Proc. IRE **37**, 10–21 (1949)

39. Z. Šidák, Rectangular confidence regions for the means of multivariate normal distributions. J. Am. Stat. Assoc. **62**, 626–633 (1967)

40. D. Slepian, The one-sided barrier problem for Gaussian noise. Bell Syst. Tech. J. **41**, 463–501 (1962)

41. M. Steiner, The strong simplex conjecture is false. IEEE Trans. Inf. Theory **40**(3), 721–731 (1994)

42. V.N. Sudakov, Geometric problems of the theory of infinite-dimensional probability distributions. Tr. Mat. Inst. Steklov **141**, 3–191 (1976)

43. Y. Sun, Stochastic iterative algorithms for signal set design for Gaussian channels and optimality of the $L2$ signal set. IEEE Trans. Inf. Theory **43**(5), 1574–1587 (1997)
44. R.M. Tanner, Contributions to the simplex code conjecture. Tech. Report No. 6151-8, Information Systems Lab., Stanford University (1970)
45. R.M. Tanner, Some content maximizing properties of the regular simplex. Pac. J. Math. **52**, 611–616 (1974)
46. C.L. Weber, *Elements of Detection and Signal Design* (McGraw-Hill, New York, 1968)

Discrete Centro-Affine Curvature for Convex Polygons

Alina Stancu

Abstract We propose a notion of centro-affine curvature for planar, convex polygons which serves to define a non-trivial affine length, and p-affine length respectively, for polygons. These concepts of affine length are shown to be similar to their counterparts defined for smooth convex bodies in that they satisfy analogous affine isoperimetric type inequalities.

1 Introduction

The study of geometries invariant under group transformations has been credited in large measure to the Erlangen program proposed by Felix Klein in 1872. Affine geometry defined by invariance under affine transformations provided a geometry less rigid than Euclidean geometry and has been particularly stimulated by the Erlangen program. Shortly afterwards, affine geometry, particularly affine differential geometry, ramified even further into equi-affine geometry, invariant under linear transformations of the ambient Euclidean space, and centro-affine geometry, invariant under linear transformations of determinant one of the ambient Euclidean space.

The seminal notion of affine surface area of convex bodies originated by Blaschke [3] is common to all three affine geometries above and, following Lutwak's extension of the Brunn-Minkowski theory to the Brunn-Minkowski-Firey theory [19, 20], its generalization to p-affine surface area is the object of high level work investigating its applications, [5, 6, 10, 11, 15, 17, 21, 22, 24, 31, 33, 34]. These applications proved to be reaching far from the initial context in which the affine and p-affine surface area arose, displaying connections to information theory and stochastic geometry, see for example [2, 8, 16, 25].

A. Stancu (✉)

Department of Mathematics and Statistics, Concordia University, Montreal, QC, Canada

e-mail: alina.stancu@concordia.ca

© Springer International Publishing AG 2018

G. Bianchi et al. (eds.), *Analytic Aspects of Convexity*, Springer INdAM Series 25,

https://doi.org/10.1007/978-3-319-71834-7_6

The invariance of geometric properties is, in fact, a crucial factor in many areas of applied mathematics, particularly in image processing and computer graphics. The affine geometry is occupying a significant place in this fields providing a more general set up then the Euclidean geometry yet simpler than projective geometry both from the analytical and computational point of view, [29].

In computer imaging and graphics, the process is often carried over to a discrete model which approximates the continuous case. Various notions of discrete curvature were thus introduced with definitions that can differ significantly, depending on the goal and the nature of the problem. It is, in fact, a specific problem that motivated us too to seek a notion of discrete centro-affine curvature. It is well known that the classical notion of affine surface area, and its p generalizations, is zero on all polytopes. In contrast, so many affine or equi-affine invariant inequalities have extremals, or conjectured extremals, on polytopes at one end, and ellipsoids at the other. However, the usual tools used for smooth convex bodies cannot carry over to convex polytopes, especially if involving affine surface area.

As mentioned before, affine surface area is one of the most powerful tools in affine differential geometry with connections in information theory and approximation theory. What was not mentioned is that, in the smooth context, the affine surface area, as well as its generalizations of p-affine surface areas, of convex bodies can be viewed as averages of the centro-affine curvature with respect to their cone-volume measure. However, the centro-affine curvature for polytopes is zero a.e. with respect to the cone volume measure, where the cone-volume measure is a notion that received recently renewed attention due to the logarithmic Minkoswki problem [4].

Discrete affine invariants for polytopes are not necessarily novel, see for example [7, 35], and elements of affine discretization appeared even earlier in [2] and [8], yet the view we adopt to define a discrete centro-affine curvature is entirely new. We use the fact that $SL(n)$ transformations preserve volumes to define a discrete centro-affine curvature for polytopes that uses both the polytope and its polar and, with its aid, we introduce a non-trivial notion of affine surface area and p-affine surface area for polytopes.

In this paper, we present the planar, thus lower dimensional case, in which affine surface area is simply affine length, yet we would like to point out that many of the results we state hold in all dimensions, though sometimes under an extra assumption on the polytopes. To be more precise, passing from $n = 2$ to $n \geq 3$, we often require that the polytope is simple, or equivalently, that its polar is simplicial. Since simple (and simplicial) polytopes are dense in the class of all polytopes in the Hausdorff distance [9], this may not be so restrictive as it seems.

The paper is structured as follows. In the next section, we define the discrete centro-affine curvature on convex, planar polytopes, followed, in Sect. 3, by a new notion of affine length and, respectively, p-affine length for polytopes, with some of their properties. We continue, in the last section, with some applications of these newly introduced concepts.

2 Discrete Centro-Affine Curvature

Let $\mathcal{P}_0^2$ be the set of polygonal convex bodies in $\mathbb{R}^2$ whose interior contain the origin. If we refer to P as the convex hull of its vertices $v_1, \ldots, v_r$, we will use the notation $[v_1, \ldots, v_r]$. Furthermore F_i, $i = 1, \ldots, r$, will denote the faces of P and $\mathcal{N} := \{u_i\}_{i=1,\ldots,r} \subset \mathbb{S}^1$ the set of unit normals to the faces of P. Alternately, we may refer to a polygon P as the intersection of the closed halfspaces bounded by the affine subspaces of $\mathbb{R}^2$ determined by F_i's, namely, $P = \bigcap_{i=1}^n \{x \in \mathbb{R}^2 \mid x \cdot u_i \le h_i, \ i = 1, \ldots, r\}$, where $h_i = dist(0, F_i)$, the dot $\cdot$ represents the usual scalar product in $\mathbb{R}^2$, and *dist* stands for the Euclidean distance in the plane.

Given $P \in \mathcal{P}_0^2$, let $P^\circ \in \mathcal{P}_0^2$ be its polar body $P^\circ = \{y \in \mathbb{R}^2 \mid x \cdot y \le 1, \text{ for all } x \in P\}$. The vertices of P° will be denoted by v_i°, $i = 1, \ldots r$, and the faces of P° will be denoted by F_i°, $i = 1, \ldots, r$. In general, we will use the convention that the quantities associated to the polar body will carry the $\circ$ notation. Furthermore, we will use the notation $\mathcal{V} := \{u_j\}_{j=1,\ldots,r} \subset \mathbb{S}^1$ for the set of directions of the vertices of P, namely $u_j = v_j / \|v_j\|$, $j = 1, \ldots, r$. With this, note that $\mathcal{N}^\circ = \mathcal{V}$ and $\mathcal{V}^\circ = \mathcal{N}$.

We mention a few additional notations: $[0, F_i]$ denotes the convex hull of the origin and the i-th side of P; the area of P, or any other compact set of $\mathbb{R}^2$, is the 2-dimensional Lebesgue measure of the compact set as a subset of $\mathbb{R}^2$ and is denoted by $A(P)$, while $|F|$ represents length as the 1-dimensional Lebsgue measure of a 1-dimensional compact set F, usually the side of a polygon.

Let $u = u_j \in \mathcal{N}$, for some $j = 1, \ldots, r$, be a fixed normal and let $\epsilon > 0$ be a relatively small number. Consider the polygonal convex body

$$P_{\epsilon,u} := \bigcap_{i=1}^r \{x \in \mathbb{R}^2 \mid \text{If } i = j: \ x \cdot u_j \le h_j + \epsilon, \text{ and if } i \ne j: \ x \cdot u_i \le h_i\}.$$

If $u \in \mathbb{S}^1 \setminus \mathcal{N}$ is a fixed normal, then u belongs to the outer normal cone of one of the vertices of P, say v_j for some $j = 1, \ldots, r$. For such a vector u, and for any $\epsilon > 0$, a positive number, let

$$P_{\epsilon,u} := [x_1, \ldots, x_j + u\epsilon, \ldots, x_r].$$

Definition 2.1 We define the centro-affine curvature of P as a function $\mathcal{K}_0(., P) : \mathbb{S}^1 \to \mathbb{R}$, by

$$\mathcal{K}_0(u, P) = \lim_{\epsilon \to 0^+} \frac{|A(P_{\epsilon,u}^\circ) - A(P^\circ)|}{A(P_{\epsilon,u}) - A(P)}, \tag{2.1}$$

where $P_{\epsilon,u}^\circ$ is the polar of $P_{\epsilon,u}$.

Theorem 2.1 *For any $P \in \mathcal{P}_0^2$, and any $u \in \mathbb{S}^1$, the centro-affine curvature of P in the direction u exists, it is finite, positive and is invariant under linear transformations of determinant one in the sense that, if $A \in SL(2)$, then $\mathcal{K}_0(Au, AP) = \mathcal{K}_0(u, P)$, for any $u \in \mathbb{S}^1$.*

Moreover, if $u = u_j \in \mathcal{N}$, for some $j = 1, \ldots, r$, then the centro-affine curvature of P in the direction u_j is the ratio between the average of areas of the triangles in P° that have v_j° as vertex and the area of the triangle formed by the origin and the j-th face of P of outer normal u_j:

$$\mathcal{K}_0(u_j, P) = \frac{\frac{1}{2}A([0, v_{j-1}^\circ, v_j^\circ, v_{j+1}^\circ])}{A([0, F_j])}. \tag{2.2}$$

Proof Let first $u = u_i$ be the outer normal of one of the polygon's sides, say F_i, for some $i = 1, \ldots, r$. Then, an easy calculation shows that

$$A(P_{\epsilon,u}) - A(P) = \frac{\epsilon}{2} (l_i + (l_i - \epsilon \cot \alpha_{i-1} - \epsilon \cot \alpha_i)), \tag{2.3}$$

where α_{i-1} and α_i are determined by the vectors u_{i-1}, u_i and u_i, u_{i+1} respectively as follows

$$\alpha_{i-1} = \arccos(u_{i-1} \cdot u_i), \quad \alpha_i = \arccos(u_i \cdot u_{i+1}), \tag{2.4}$$

where the index by i is taken modulo r.

As the i-th side of P is translated, the i-th vertex of P° moves inward from the position $\frac{1}{h_i} u_i$ to $\frac{1}{h_i + \epsilon} u_i$. Consequently,

$$|A(P_{\epsilon,u}^\circ) - A(P^\circ)| = \frac{\epsilon}{2} \left(\frac{\sin \alpha_{i-1}}{h_{i-1} h_i (h_i + \epsilon)} + \frac{\sin \alpha_i}{h_{i+1} h_i (h_i + \epsilon)} \right), \tag{2.5}$$

The explicit formulas for $A(P_{\epsilon,u}) - A(P)$ and $|A(P_{\epsilon,u}^\circ) - A(P^\circ)|$ prove the existence of the limit for directions normal to the sides of P, as well as formula (2.2).

If $u \in \mathcal{V}$, say $u = \frac{v_i}{\|v_i\|}$ for some i, we note a reversal in the type of deformation that P and P° undergo. The i-th vertex of P moves farther out by ϵ in the direction u, while the polar P° will have its i-face move parallel inside. Thus, almost identical calculations with the ones above give

$$A(P_{\epsilon,u}) - A(P) = \frac{\epsilon}{2} \left(\frac{\sin \alpha_{i-1}^\circ}{h_{i-1}^\circ h_i^\circ (h_i^\circ + \epsilon)} + \frac{\sin \alpha_i^\circ}{h_{i+1}^\circ h_i^\circ (h_i^\circ + \epsilon)} \right) \tag{2.6}$$

and

$$|A(P_{\epsilon,u}^\circ) - A(P^\circ)| = \frac{\epsilon}{2} \left(l_i^\circ + (l_i^\circ + \epsilon \cot \alpha_{i-1}^\circ + \epsilon \cot \alpha_i^\circ) \right), \tag{2.7}$$

where α_{i-1}° and α_i° are determined by the vectors u_{i-1}°, u_i° and u_i°, u_{i+1}° in the analogous manner described earlier. We conclude that the limit definition of the centro-affine curvature in the direction $u \in \mathcal{V}$, $u = \frac{v_i}{||v_i||}$, is

$$\mathcal{K}_0(u, P) = \frac{\frac{1}{2}A([0, v_{i-1}, v_i, v_{i+1}]}{A([0, F_i^{\circ}])}. \tag{2.8}$$

Finally, if $u \in \mathbb{S}^1 \setminus (\mathcal{N} \cup \mathcal{V})$, then u belongs to the normal cone of some vertex i of P, of position v_i but $u \neq \frac{v_i}{||v_i||}$, and the change in area of the polygon equals the area of two triangles, one made by the $(i-1)$-th side with the vector ϵu originating at v_i, the other made by the i-th side with the vector ϵu originating at v_i:

$$A(P_{\epsilon,u}) - A(P) = \frac{\epsilon}{2}\left[l_{i-1} \sin\left(\arccos\left(\frac{f_{i-1} \cdot u}{||f_{i-1}||}\right)\right) + l_i \sin\left(\arccos\left(\frac{f_i \cdot u}{||f_i||}\right)\right)\right], \tag{2.9}$$

where the vectors f_{i-1}, f_i are defined by $f_i = v_{i+1} - v_i$ and $f_{i-1} = v_{i-1} - v_i$.

On the other hand, as the i-side of P°, at distance $h_i^{\circ}(\epsilon) = \frac{1}{||v_i + \epsilon u||}$ from the origin, changes orientation as ϵ varies in the expression of the vertex $v_{i,\epsilon} = v_i + \epsilon u$ of $P_{\epsilon,u}$. Thus, we have

$$P_{\epsilon,u}^{\circ} = \bigcap_{i=1}^{n}\left\{x \in \mathbb{R}^2 \mid x \cdot \frac{v_i + \epsilon u}{||v_i + \epsilon u||} \leq \frac{1}{||v_i + \epsilon u||} \text{ and if } j \neq i: x \cdot v_j \leq \frac{1}{||v_j||}\right\}. \tag{2.10}$$

Denoting by $\beta_i(\epsilon) = \arccos\left(\frac{v_i \cdot (v_i + \epsilon u)}{||v_i|| \cdot ||v_i + \epsilon u||}\right)$ and by $l_{i,j,\epsilon}^{\circ}$, $l_{i,j}^{\circ}(\epsilon), j = 1, 2$, the lengths of the segments obtained on F_i° and, respectively, on $F_{i,\epsilon}^{\circ}$, of normals v_i respectively $v_i + \epsilon u$, we have:

$$|A(P_{\epsilon,u}^{\circ}) - A(P^{\circ})| = \frac{1}{2} l_{i,1,\epsilon}^{\circ} l_{i,1}^{\circ}(\epsilon) \sin(\beta_i(\epsilon)) + \frac{1}{2} l_{i,2,\epsilon}^{\circ} l_{i,2}^{\circ}(\epsilon) \sin(\beta_i(\epsilon)). \tag{2.11}$$

On the other hand,

$$\lim_{\epsilon \to 0^+} l_{i,j,\epsilon}^{\circ} = \lim_{\epsilon \to 0^+} l_{i,j}^{\circ}(\epsilon) = \lambda_j l_i^{\circ}, j = 1, 2,$$

for some $\lambda_j \in [0, 1], j = 1, 2, \lambda_1 + \lambda_2 = 1$, and

$$\lim_{\epsilon \to 0^+} \frac{\sin(\beta_i(\epsilon))}{\epsilon} = \lim_{\epsilon \to 0^+} \frac{1}{\epsilon} \sin\left[\arccos\left(\frac{||v_i||^2 + \epsilon v_i \cdot u}{||v_i|| \cdot ||v_i + \epsilon u||}\right)\right]$$

$$= \frac{\sqrt{||v_i||^2 - (v_i \cdot u)^2}}{||v_i||^2}.$$

The invariance of the centro-affine curvature under linear transformations of determinant 1 is immediate from its definition as the areas of P, $P_{\epsilon,u}$, and their polars, do not change after the transformation is applied.   □

As we noted in the proof of the previous theorem, whether the deformation of P is done toward its exterior or toward the interior of the polygon, the limit does not change. This is made more precise by the statement of the next proposition.

Let $u = u_j \in \mathcal{N}$, for some $j = 1, \ldots, r$, be a fixed normal and let $\epsilon > 0$ be a relatively small number. The size of ϵ is now somewhat relevant in the sense of keeping the convexity of the new set and the assumption that the origin is included in its interior. Consider the polygonal convex body

$$P_{-\epsilon,u} := \bigcap_{i=1}^{r} \{x \in \mathbb{R}^2 \mid \text{If } i = j : \ x \cdot u_j \leq h_j - \epsilon, \text{ and if } i \neq j : \ x \cdot u_i \leq h_i\}.$$

If $u \in \mathbb{S}^2 \setminus \mathcal{N}$ is a fixed normal, then u belongs to the outer normal cone of one of the vertices of P denoted by v_i, say v_j for some $j = 1, \ldots, r$. For such a vector u, and for any $\epsilon > 0$, a positive number, let

$$P_{\epsilon,u} := [v_1, \ldots, v_j - u\epsilon, \ldots, v_r].$$

Proposition 2.1 *For any* $P \in \mathcal{P}_0^2$, *and for any* $u \in \mathbb{S}^1$, *the centro-affine curvature of* P *in the direction* u *satisfies*

$$\mathcal{K}_0(u, P) = \lim_{\epsilon \to 0^+} \frac{A(P_{-\epsilon,u}^\circ) - A(P^\circ)}{|A(P_{-\epsilon,u}) - A(P)|}, \tag{2.12}$$

More generally, let $\bar{\epsilon}$ *be a small real number for which* $P_{\bar{\epsilon},u} = P_{|\epsilon|,u}$ *if* $\bar{\epsilon} > 0$ *and* $P_{\bar{\epsilon},u} = P_{-|\epsilon|,u}$ *if* $\bar{\epsilon} < 0$. *Then*

$$\mathcal{K}_0(u, P) = \lim_{\bar{\epsilon} \to 0} \frac{|A(P_{\bar{\epsilon},u}^\circ) - A(P^\circ)|}{|A(P_{\bar{\epsilon},u}) - A(P)|}, \tag{2.13}$$

A second remark coming from the previous calculations is that, if $u_j \in \mathcal{V}$, then evaluating the limit in (2.1) is equivalent to the reciprocal of the calculations done now for P° and the direction to its j-face F_i°, thus $u_j \in \mathcal{N}^\circ$, for negative ϵ in the sense just discussed above. More precisely,

Corollary 2.1 *Let* P *be a convex polygonal body in* $\mathbb{R}^2$ *containing the origin in its interior and let* P° *denote its polar convex body. Then, for any unit vector* u *of* $\mathbb{R}^2$, *we have*

$$\mathcal{K}_0(u, P) \cdot \mathcal{K}_0(u, P^\circ) = 1. \tag{2.14}$$

This result is analogous to the reciprocality result for the centro-affine curvature of a sufficiently regular convex body and that of its polar as obtained by Salkowski [28], in dimension three, then Kaltenbach [13] in all dimensions, and later generalized by Hug, [12].

Theorem 2.2 *Let* $\gamma : [0, L] \to \mathbb{R}^2$ *be a* C_+^2 *convex curve in the plane parameterized by arclength s whose support function is h and curvature function is, respectively, k.*
Suppose that $\{x_0 = \gamma(0), x_1 = \gamma(\Delta s), x_2 = \gamma(2\Delta s), \dots, x_{n+1} = \gamma(n\Delta s) = \gamma(L)\}$ *is a partition of the curve with* $0 < \Delta s << L$, *a small positive real number. Let* $P_{\Delta s}$ *be the polygonal convex body* $P_{\Delta s} := [\gamma(0), \gamma(\Delta s), \dots, \gamma((n-1)\Delta s)]$ *and let* $u(s) \in \mathcal{N}(P_{\Delta s})$.
If $u = \lim_{\Delta s \to 0+} u(s)$, *then*

$$\lim_{\Delta s \to 0+} \mathcal{K}_0(u(s), P_{\Delta s}) = \frac{k}{h^3}(v^{-1}(u)), \tag{2.15}$$

where $v : [0, L] \to \mathbb{S}^1$ *is the Gauss map from the* $\gamma(s)$ *to the unit outer normal to* γ *at* $\gamma(s)$.

Proof Denote by $\Delta \theta_i$ the angle between the outer normals of two consecutive, $i, i+1$, faces of $P_{\Delta s}$. Then, using (2.2), we have for very small Δs:

$$\mathcal{K}_0(u(s), P_{\Delta s}) = \frac{\frac{1}{2} \cdot \frac{1}{2h_i} \left(\frac{\sin \Delta \theta_i}{h_{i-1}} + \frac{\sin \Delta \theta_{i+1}}{h_{i+1}} \right)}{\frac{1}{2} h_i l_i} \approx \frac{\frac{1}{2} \cdot \frac{1}{2h_i} \left(\frac{\sin \Delta \theta_i}{h_{i-1}} + \frac{\sin \Delta \theta_{i+1}}{h_{i+1}} \right)}{\frac{1}{2} h_i \Delta s}$$

$$\approx \frac{\frac{1}{2}(h_{i+1} \Delta \theta_i + h_i \Delta \theta_{i+1})}{h_{i-1} h_i^2 h_{i+1} \Delta s} \to \frac{\frac{d\theta}{ds}}{h^3}(v^{-1}(u)) = \frac{k}{h^3}(v^{-1}(u)). \qquad \square$$

3 Discrete Affine Length

We will now use the notion of centro-affine curvature for polygons introduced earlier to define a new notion of p-affine length for a polygonal convex curve. It is known that the classical definition of p-affine length defined for sufficiently regular curves, or boundaries of convex bodies in $\mathbb{R}^2$, is zero for all polygonal convex bodies. Here we propose a non-trivial notion of affine length for polygonal convex bodies that will have similar properties with the classical notion when the latter is restricted to smooth convex bodies.

Definition 3.1 For any real number $p \neq -2$, let the (discrete) p-affine length of a polygon P be

$$\tilde{\Omega}_p(P) = \int_{\mathbb{S}^1} \mathcal{K}_0(u, P)^{\frac{p}{p+2}} \, dA_P(u), \tag{3.1}$$

where dA_P is the cone volume measure of P. For $p = 1$, we will call the (discrete) p-affine length, the (discrete) affine length.

For simplicity, we will omit the word discrete thereafter if there is no risk of confusing the reader.

A few words are in order regarding the cone volume measure and the integrability of the function $u \mapsto \mathcal{K}_0(u, P)^{\frac{p}{p+2}}$ with respect to the cone volume measure. The cone volume measure is a classical notion made famous by Böröczky, Lutwak, Yang and Zhang in what is now known as the L_0-Minkowski problem, [4]. We will keep the name of cone *volume* measure even if in, the plane, the word *area* would be more appropriate. If K is any convex body in $\mathbb{R}^2$, the cone volume measure of K, is a measure on $\mathbb{S}^1$, defined by

$$dA_K(u) = \frac{1}{2} h_K(u) \, dL_K(u), \tag{3.2}$$

where h is the support function of K as a function on u unit vector identified with a point on the unit sphere, and dL_K is the length (Hausdorff) measure of the boundary of K. In other references, the cone volume measure does not have the normalization of $\frac{1}{2}$ that we consider so that $\int_{\mathbb{S}^1} dA_K = A(K)$.

In the case of a polygonal convex body, the length measure is concentrated on the finite number of unitary directions normal to the sides of the polygon and, for each such vector u_i, the measure equals the length of the side L_i with u_i as outer normal. Therefore, the cone volume measure of a polygon is also concentrated along the same finite number of unitary directions and, for each such direction u_i, the measure equals the area of $[0, F_i]$.

Moreover, as the support of the cone volume measure of a fixed convex planar polygon consists of a finite number of unitary directions, and $\mathcal{K}_0(u, P)$ is strictly positive and finite in those directions, $\mathcal{K}_0(u, P)^{\frac{p}{p+2}}$ is integrable with respect to the cone volume measure of the polygon. Consequently, the discrete centro-affine p-length is well-defined and is a strictly positive, finite real value.

The next two propositions summarize some properties of the discrete centro-affine p-length that have familiar counterparts in the classical case of the centro-affine p-length for smooth planar convex bodies.

Proposition 3.1

(i) $\tilde{\Omega}_p(P)$ is a global SL(2) invariant of P.
(ii) $\tilde{\Omega}_0(P) = A(P)$ and $\tilde{\Omega}_\infty(P) = \lim_{p \to \infty} \tilde{\Omega}_p(P) = A(P^\circ)$.
(iii) $\tilde{\Omega}_p(P)$ is positively homogenous of degree $\frac{2(2-p)}{2+p}$, $p \neq -2$, i.e.

$$\tilde{\Omega}_p(\lambda P) = \lambda^{\frac{2(2-p)}{2+p}} \tilde{\Omega}_p(P), \quad for \ \lambda > 0.$$

Proof

(i) This fact is a direct consequence of the fact that both the discrete centro-affine curvature of P (particularly when restricted to the directions of the outer normals to the sides) and the cone volume measure are invariant under linear transformations of determinant 1.

(ii) When $p = 0$, the definition (3.1) is, trivially,

$$\tilde{\Omega}_0(P) = \int_{\mathbb{S}^1} dA_P(u) = \sum_{i=1}^{r} A([0, F_i]) = A(P).$$

As mentioned before, for each $u \in \mathcal{N}$, the centro-affine curvature of P in the direction of u, $\mathcal{K}_0(u, P)$, is a strictly positive, finite real number. Let g be the constant function on $\mathcal{N}$ equal to the maximum after u of $\mathcal{K}_0(u, P)$. The function g is integrable with respect to the cone volume measure of P. Now, as $p \to \infty$, the fraction $\frac{p}{p+2} \to 1$ and $\mathcal{K}_0(u, P)^{\frac{p}{p+2}}$ converges pointwise to $\mathcal{K}_0(u, P)$. Thus, for any sequence of real numbers $\{p_l\}_l$ diverging to infinity, by Lebesgue's dominated convergence theorem,

$$\lim_{l \to \infty} \tilde{\Omega}_{p_l}(P) = \sum_{i=1}^{r} \frac{\frac{1}{2} A([0, v_{j-1}^\circ, v_j^\circ, v_{j+1}^\circ])}{A([0, F_j])} A([0, F_i]) = A(P^\circ), \qquad (3.3)$$

which, as the area of each triangle $A([0, F_j^\circ]$ is counted twice, concludes the proof.

(iii) Easy to check directly. $\qquad\qquad\qquad\qquad\qquad\qquad\qquad\qquad\qquad\qquad\qquad\qquad\square$

For the next result, recall that the Santaló point of P, $s(P)$, is the point x of P for which $A(P) \cdot A((P - x)^\circ)$ is minimal. It is known that this point exists and is unique, [30].

Proposition 3.2

(i) If $p > 0$, or $p < -2$, then

$$\frac{\tilde{\Omega}_p(P)^{2+p}}{Vol(P)^{2-p}} \leq (A(P) \cdot A(P^\circ))^p, \qquad (3.4)$$

with equality if and only if P is a convex polygon whose restriction of the centro-affine curvature function to $\mathcal{N}$ is constant. Moreover, if the origin is the Santaló point of P, then

$$\frac{\tilde{\Omega}_p(P)^{2+p}}{A(P)^{2-p}} < \omega_2^{2p}, \qquad (3.5)$$

where ω_2 is the area of the Euclidean unit ball in $\mathbb{R}^2$.

(ii) If $-2 < p < 0$, *then*

$$\frac{\tilde{\Omega}_p(P)^{2+p}}{A(P)^{2-p}} \geq (A(P) \cdot A(P^\circ))^p, \tag{3.6}$$

with equality if and only if P is a convex polygon whose restriction of the centro-affine curvature function to $\mathcal{N}$ *is constant.*

Proof The first and third inequality are derived directly from Hölder's inequality applied to $\tilde{\Omega}_p(P)$ as follows.

Since

$$\int_{\mathbb{S}^1} \mathcal{K}_0(u, P)^{\frac{p}{p+2}} \, dA_P(u) = \sum_{i=1}^{r} \mathcal{K}_0(u_i, P)^{\frac{p}{p+2}} A([0, F_i]) \tag{3.7}$$

where u_i, $i = 1, \ldots, n$, are the vectors of $\mathcal{N}$ on which the cone volume measure is concentrated, note that, for $p/(p + 2) > 0$, we have

$$\left[\sum_{i=1}^{r} \mathcal{K}_0(u_i, P)^{\frac{p}{p+2}} A([0, F_i])\right]^{\frac{p+2}{p}} \times \left[\sum_{i=1}^{r} A([0, F_i])\right]^{-\frac{2}{p}}$$

$$\leq \sum_{i=1}^{r} \mathcal{K}_0(u_i, P) A([0, F_i]) = A(P^\circ). \tag{3.8}$$

For (3.6), Hölder's inequality is reversed. In both cases, it is obvious that the equalities occur if and only if P has constant discrete centro-affine curvature on $\mathcal{N}$.

The inequality (3.5) is a consequence of Blaschke-Santaló's inequality which reaches its extremal for ellipsoids and, consequently, this shows that the inequality (3.4) is actually strict for all polygons. The value ω_2^{2p} is the supremum after all polygons of the (discrete) centro-affine quotient obtained for any sequence of polygons converging in the Hausdorff metric to an ellipsoid.

It is interesting to note that, due to the Bourgain-Milman constant, the isoperimetric ratio for $-2 < p < 0$ is bounded away from zero. Connections with the Mahler conjecture, known to be true in $\mathbb{R}^2$, can also be exploited here to provide a lower bound and the equality cases for it, [23]. $\qquad\square$

The following proposition corresponds to a result for the classical affine surface area settled by Werner-Ye, [33].

Proposition 3.3 *The function*

$$p \to \left(\frac{\tilde{\Omega}_p(P)}{A(P)}\right)^{\frac{2+p}{p}}$$

defined for $p \in (0, +\infty)$ *(or, respectively, on* $(-\infty, -2)$ *or* $(-2, 0)$*) is increasing in* p.

Proof If $0 < p < q$, the result comes directly from Hölder's inequality:

$$\tilde{\Omega}_p(P) = \int_{\mathbb{S}^1} \mathcal{K}_0(u, P)^{\frac{p}{p+2}} \, dA_P(u) = \int_{\mathbb{S}^1} \left(\mathcal{K}_0(u, P)^{\frac{q}{q+2}} \right)^{\frac{p(2+q)}{q(2+p)}} \, dA_P(u)$$

$$\leq \tilde{\Omega}_p(P)^{\frac{p(2+q)}{q(2+p)}} A(P)^{\frac{(q-p)2}{(2+p)q}}.$$

The other cases are similar. □

4 Applications

4.1 Applications to Information Theory

In this section, we will present some results about certain other constructions of global $SL(n)$-invariants for polytopes that carry over properties from the smooth case.

In a manner similar to [17], any functional of the form

$$\tilde{\Omega}_f(P) := \int_{\mathbb{S}^1} f(\mathcal{K}_0) \, dA_P(u), \tag{4.1}$$

where $f : [0, \infty) \to [0, \infty)$ is a positive continuous function, is a global $SL(2)$-invariant of P. Two such invariants stand out in this class, not in the least because of the analogy with the smooth case where they are, each, related to an information theory-type inequality, [25] and [32].

Definition 4.1 Let P be a convex polygon in $\mathbb{R}^2$ containing the origin in its interior. Then the $SL(2)$-invariants $\tilde{\Omega}_P$ and $\tilde{\Lambda}_P$ are defined as follows:

$$\tilde{\Omega}_P := \lim_{p \to +\infty} \left(\frac{\tilde{\Omega}_p(P)}{A(P^\circ)} \right)^{2+p} \tag{4.2}$$

and, respectively,

$$\tilde{\Lambda}_P := \lim_{p \to +\infty} \left(\frac{\tilde{\Omega}_{-\frac{2}{2p}}(P)}{A(P)} \right)^{2^p}. \tag{4.3}$$

Proposition 4.1 *For any $P \in \mathcal{P}_0^2$, we have*

$$\tilde{\Omega}_P = \exp\left[-\frac{2}{A(P^\circ)} \int_{\mathbb{S}^1} \mathcal{K}_0(u, P) \ln \mathcal{K}_0(u, P)\, dA_P(u)\right], \qquad (4.4)$$

and

$$\tilde{\Lambda}_P = \exp\left[\frac{1}{A(P)} \int_{\mathbb{S}^1} \ln \mathcal{K}_0(u, P)\, dA_P(u)\right]. \qquad (4.5)$$

Moreover, $\tilde{\Omega}_P$ and $\tilde{\Lambda}_P$ satisfy the following information theory-type inequalities, namely

$$\tilde{\Lambda}^{-2}(P) \geq \left[\frac{A(P)}{A(P^\circ)}\right]^2 \geq \tilde{\Omega}_P. \qquad (4.6)$$

Equality occurs above if and only if the centro-affine curvature of P, as a function on $\mathcal{N}$, is constant.

Proof The proof of (4.4) is a consequence of l'Hôpital's theorem:

$$\exp\left[\lim_{p \to +\infty} \frac{\ln\left(\frac{\tilde{\Omega}_p(P)}{A(P^\circ)}\right)}{\frac{1}{2+p}}\right] = \exp\left[\lim_{p \to +\infty} \frac{\frac{d}{dp}\tilde{\Omega}_p(P)}{-\frac{1}{(2+p)^2}}\right]$$

$$= \exp\left[-2 \lim_{p \to +\infty} \frac{\int_{\mathbb{S}^1} \mathcal{K}_0(u, P)^{\frac{p}{p+2}} \ln \mathcal{K}_0(u, P)\, dA_P(u)}{\tilde{\Omega}_p(P)}\right]$$

$$= \exp\left[-\frac{2}{A(P^\circ)} \int_{\mathbb{S}^1} \mathcal{K}_0(u, P) \ln \mathcal{K}_0(u, P)\, dA_P(u)\right].$$

It is precisely this expression that is related to Kullback-Leibler divergence. The Kullback-Leibler divergence D_{KL} of two specific probability measures Q_1, Q_2 on the boundary of P is defined as $D_{KL}(Q_1 \| Q_2) = \int_{\partial P} \ln \frac{dQ_1}{dQ_2}\, dQ_1$, where $\frac{dQ_1}{dQ_2}$ is the Radon-Nikodym derivative of Q_1 with respect to Q_2, provided that the right-hand side exists. In this case, $dQ_1(u) = \frac{1}{A(P^\circ)} \mathcal{K}_0(u, P)\, dA_P(u)$ while $dQ_1(u) = \frac{1}{A(P)}\, dA_P(u)$. Thus

$$D_{KL}(Q_1 \| Q_2) = -\frac{1}{2} \ln \tilde{\Omega}_P + \ln \frac{A(P)}{A(P^\circ)}. \qquad (4.7)$$

The proof of (4.5) follows the same principle of [32] as in the smooth case and is also a consequence of l'Hôpital's theorem.

To prove that $\left[\frac{A(P)}{A(P^\circ)}\right]^2 \geq \tilde{\Omega}_P$, note that, for all positive p,

$$
\begin{aligned}
\left(\frac{\tilde{\Omega}_p(P)}{A(P^\circ)}\right)^{2+p} &= \left(\frac{\tilde{\Omega}_p(P)}{A(P)} \cdot \frac{A(P)}{A(P^\circ)}\right)^{2+p} \\
&= \left[\left(\frac{\tilde{\Omega}_p(P)}{A(P)}\right)^{\frac{2+p}{p}}\right]^p \cdot \left(\frac{A(P)}{A(P^\circ)}\right)^{2+p} \\
&\leq \left(\frac{A(P^\circ)}{A(P)}\right)^p \cdot \left(\frac{A(P)}{A(P^\circ)}\right)^{2+p} \\
&\leq \left(\frac{A(P)}{A(P^\circ)}\right)^2,
\end{aligned}
\tag{4.8}
$$

where $\left(\frac{\tilde{\Omega}_p(P)}{A(P)}\right)^{\frac{2+p}{p}} \leq \frac{A(P^\circ)}{A(P)}$ comes from the fact that for any $0 < p < q$, we have $\left(\frac{\tilde{\Omega}_p(P)}{A(P)}\right)^{\frac{2+p}{p}} \leq \left(\frac{\tilde{\Omega}_q(P)}{A(P)}\right)^{\frac{2+q}{q}}$, by Proposition 3.3, and we let q go to infinity.

The last inequality, together with (4.7), implies that $D_{KL}(Q_1 \| Q_2) \geq 0$, a result known in information theory as Gibbs' inequality. Note that the equality holds if the centro-affine curvature of P is constant along $\mathcal{N}$ and thus $Q_1 = Q_2$ on $\mathcal{N}$.

To complete the proof of (4.6), we will use a generalized Hölder's inequality that we have encountered in [1] in a slightly different context. If $d\omega$ is a volume form on $\mathbb{S}^1$, g is a positive function on $\mathbb{S}^1$ and F is a increasing real, positive function, then

$$
\frac{\int_{\mathbb{S}^1} gF(g)\, d\omega}{\int_{\mathbb{S}^1} F(g)\, d\omega} \geq \frac{\int_{\mathbb{S}^1} g\, d\omega}{\int_{\mathbb{S}^1} d\omega}.
\tag{4.9}
$$

If F is strictly increasing, then equality occurs if and only if g is a constant function. As the logarithm function is strictly increasing on $(0, \infty)$, we have

$$
\frac{\int_{\mathbb{S}^1} \mathcal{K}_0(u, P) \ln \mathcal{K}_0(u, P)\, dA_P(u)}{\int_{\mathbb{S}^1} \ln \mathcal{K}_0(u, P)\, dA_P(u)} - \frac{\int_{\mathbb{S}^1} \mathcal{K}_0(u, P)\, dA_P(u)}{\int_{\mathbb{S}^1} dA_P(u)} \geq 0,
\tag{4.10}
$$

which, after re-arranging terms, is equivalent to $\ln \tilde{\Omega}_P^{-\frac{1}{2}} \geq \ln \tilde{\Lambda}_P$. Finally, from Jensen's inequality,

$$
\ln \tilde{\Lambda}_P = \int_{\mathbb{S}^1} \ln \mathcal{K}_0(u, P) \frac{dA_P(u)}{A(P)} \leq \ln \frac{A(P^\circ)}{A(P)},
\tag{4.11}
$$

and both equality cases follow immediately. □

Applying the previous proposition to both P, and its polar, we obtain:

Corollary 4.1 *For any* $P \in \mathcal{P}_0^2$,

$$\tilde{\Omega}_P \cdot \tilde{\Omega}_{P^\circ} \leq 1 \quad and \quad \tilde{\Lambda}_P \cdot \tilde{\Lambda}_{P^\circ} \leq 1, \tag{4.12}$$

with equality if and only if the centro-affine curvature of P is constant along the directions of $\mathcal{N}$.

We would like to call *centro-affinely regular* a convex polygon $P \in \mathcal{P}_0^2$ whose centro-affine curvature of P is constant along the directions of $\mathcal{N}$. While we know that all linear images of regular polygons with the center at the origin are centro-affinely regular, we do not know if a centro-affinely regular polygon can be different than the linear image of a regular polygon. We, therefore, refrain from giving a formal definition for a centro-affinely regular a convex polygon until the previous question is settled.

4.2 A Connection to Geominimal Surface Area

An interesting notion of affine surface area for convex bodies K, which is independent of the boundary regularity of K, has been introduced by Petty [26]. Much later, in 1991, Lutwak was able to show certain compatibility between the geominimal surface area and affine surface area for smooth convex bodies, [18]. Consequently, certain inequalities between these notions followed, see also [27] and, for a survey, [14].

Recall that, in the planar context, the geominimal surface area for a convex body K in $\mathbb{R}^2$ defined by Petty is

$$G(K) = \inf \left\{ nV(K,T) = \int_{\mathbb{S}^1} h_T(u)\,dS_K(u) \mid T \in \mathcal{T}^2 \right\}, \tag{4.13}$$

where $\mathcal{T}^2 = \{T$ convex body in $\mathbb{R}^2, s(T) = 0, Vol(T^\circ) = \omega_2\}$, [26].

It was shown by Petty in the same paper that a convex body T realizing the infimum exists. Moreover, if K is a polytope, it is easy to see that T is a polytope with the same set of outer normals to the faces. Indeed, assuming that T is not a polytope with outer normals $\mathcal{N}$, note that, to know $G(P)$, it suffices to know the support function of T in the directions $u_i \in \mathcal{N}_P$. Thus, consider the polytope $P_T := \{x \in \mathbb{R}^2 \mid x \cdot u_i \leq h_T(u_i), \ \forall u_i \in \mathcal{N}\}$. Due to the convexity of T and P_T, we have that $T \subseteq P_T$ and thus $P_T^\circ \subseteq T^\circ$. Hence $A(P_T^\circ) < \omega_2$, with equality only if $P_T = T$. So, if the inequality is strict, we can shrink P_T by an $\eta < 1$ so that its polar has area $\frac{1}{\eta^2} A(P_T^\circ) = \omega_2$. However, then $V(P, \eta P_T) < V(P, T)$ contradicting the definition of $G(P)$.

Therefore, for a planar polygon P with support functions to the sides $h_1, \ldots, h_r$ and sides of lengths $l_1, \ldots, l_r$, the problem can be formulated as the minimization problem:

Find the polygon $T \in \mathcal{T}^2$ with $\mathcal{N}_T = \mathcal{N}_P$, whose support values to the sides $\tilde{h}_1, \ldots, \tilde{h}_r$ are defined by

$$\min \sum_{i=1}^{r} \tilde{h}_i l_i, \quad \text{with} \quad \frac{1}{2} \sum_{i=1}^{r} \frac{1}{\tilde{h}_i \tilde{h}_{i+1}} \sin \alpha_i (= Vol(T)) = \omega_2. \tag{4.14}$$

Treating this optimization problem with constraint for $(\tilde{h}_1, \ldots, \tilde{h}_r) \in \mathbb{R}^r$, the Karush-Kuhn-Tucker (KKT) conditions give:

$$l_i = \frac{\lambda}{2} \left[-\frac{1}{\tilde{h}_i^2 \tilde{h}_{i+1}} \sin \alpha_i - \frac{1}{\tilde{h}_i^2 \tilde{h}_{i-1}} \sin \alpha_{i-1} \right], \quad i = 1, \ldots, r, \tag{4.15}$$

$$\frac{1}{2} \sum_{i=1}^{r} \frac{1}{\tilde{h}_i \tilde{h}_{i+1}} \sin \alpha_i = \omega_2.$$

Thus $\tilde{h}_i l_i = |\lambda| \left(A([0, v_{j-1}^\circ, v_j^\circ, v_{j+1}^\circ]) \right)$, and furthermore

$$G(P) = 2|\lambda| A(T^\circ) = 2|\lambda| \omega_2. \tag{4.16}$$

In another form, (4.16) is

$$l_i = |\lambda| \frac{A([0, \tilde{v}_{j-1}^\circ, \tilde{v}_j^\circ, \tilde{v}_{j+1}^\circ])}{A([0, \tilde{F}_j])} \tilde{l}_i, \quad i = 1, \ldots, r. \tag{4.17}$$

Note from (4.17) that, if T is homothetic to P, or what Petty defines as P is self-minimal, then by Minkowski's existence theorem, the centro-affine curvature of T is constant on $\mathcal{N}$. Consequently, P is centro-affinely regular, in the sense discussed before with respect to its Santaló point. Thus we have proved a converse to Petty's theorem [26] which states that any affinely regular poytope is self-minimal. This converse is a particular reason for which knowing if there exist centro-affinely regular polygons that are not affinely regular would be interesting:

Proposition 4.2 *If P is self-minimal, then P is centro-affinely regular.*

Acknowledgements The author is thankful to Gabriele Bianchi, Andrea Colesanti, Paolo Gronchi, the organizers of the 2016 Workshop on Analytic Aspects of Convexity, as well as to Stefano Campi for the invitation to participate, to the Istituto Nazionale di Alta Matematica (INdAM) for its hospitality and, to all of the above, for the stimulating atmosphere during her stay there.

This work was partially supported by an NSERC grant.

References

1. B. Andrews, Evolving convex curves. Calc. Var. Partial Differ. Equ. **7**, 315–371 (1998)
2. I. Bárány, Affine perimeter and limit shape. J. Reine Angew. Math. **484**, 71–84 (1997)
3. W. Blaschke, Über affine Geometrie XXIX: affinminimalflächen. Math. Z. **12**, 262–273 (1922)
4. K. Böröczky, E. Lutwak, D. Yang, G. Zhang, The logarithmic Minkowski problem. J. Am. Math. Soc. **26**, 831–852 (2012)
5. K. Chou, X.-J. Wang, The L_p-Minkowski problem and the Minkowski problem in centroaffine geometry. Adv. Math. **205**, 33–83 (2006)
6. A. Cianchi, E. Lutwak, D. Yang, G. Zhang, Affine moser-trudinger and morrey-sobolev inequalities. Calc. Var. Partial Differ. Equ. **36**, 419–436 (2009)
7. M. Craizer, R.C. Teixeira, M.A.H.B. da Silva, Affine properties of convex equal-area polygons. Discret. Comput. Geom. **48**, 580–595 (2012)
8. P.M. Gruber, Asymptotic estimates for best and stepwise approximation of convex bodies II. Forum Math. **5**, 521–538 (1993)
9. B. Grünbaum, *Convex Polytopes Pure and Applied Mathematics*, vol. 16 (Interscience Publishers, Wiley, Inc., New York, 1967)
10. C. Haberl, F. Schuster, Asymmetric affine L_p Sobolev inequalities. J. Funct. Anal. **257**, 641–658 (2009)
11. C. Haberl, F. Schuster, General L_p affine isoperimetric inequalities. J. Differ. Geom. **83**, 1–26 (2009)
12. D. Hug, Curvature relations and affine surface area for a general convex body and its polar. Results Math. **29**, 233–248 (1996)
13. F.J. Kaltenbach, Asymptotisches verhalten zufälliger konvexer polyeder. Dissertation, Freiburg, 1990
14. K. Leichtweiss, *Affine Geometry of Convex Bodies* (Wiley, New York, 1999)
15. M. Ludwig, General affine surface areas. Adv. Math. **224**, 2346–2360 (2010)
16. M. Ludwig, Fisher information and matrix-valued valuations. Adv. Math. **226**, 2700–2711 (2011)
17. M. Ludwig, M. Reitzner, A classification of $SL(n)$ invariant valuations. Ann. Math. **172**, 1223–1271 (2010)
18. E. Lutwak, Extended affine surface area. Adv. Math. **85**, 39–68 (1991)
19. E. Lutwak, The Brunn-Minkowski-Firey theory. I: mixed volumes and the Minkowski problem. J. Differ. Geom. **38**, 131–150 (1993)
20. E. Lutwak, The Brunn-Minkowski-Firey theory II: affine and geominimal surface areas. Adv. Math. **118**, 244–294 (1996)
21. E. Lutwak, D. Yang, G. Zhang, L_p affine isoperimetric inequalities. J. Differ. Geom. **56**, 111–132 (2000)
22. E. Lutwak, D. Yang, G. Zhang, Sharp affine L_p Sobolev inequalities. J. Differ. Geom. **62**, 17–38 (2002)
23. K. Mahler, Ein Minimalproblem für konvexe Polygone. Mathematika B (Zutphen) **7**, 118–127 (1938)
24. M. Meyer, E. Werner, On the p-affine surface area. Adv. Math. **152**, 288–313 (2000)
25. G. Paouris, E. Werner, Relative entropy of cone measures and L_p centroid bodies. Proc. Lond. Math. Soc. **104**, 253–286 (2012)
26. C.M. Petty, Geominimal surface area. Geom. Dedicata **3**, 77–97 (1974)
27. C.M. Petty, Affine isoperimetric problems, in *Discrete Geometry and Convexity, Proc. Conf. New York 1982*, ed. by J.E. Goodman, E. Lutwak, J. Malkevitch, R. Pollack, Annals of the New York Academy of Sciences, vol. 440 (New York Academy of Sciences, New York, 1985), pp. 113–127
28. E. Salkowski, *Affine Differentialgeometrie. Göschens Lehrbücherei, Band 22* (Walter de Gruyter & Co., Berlin, 1934)

29. G. Sapiro, *Geometric Partial Differential Equations and Image Analysis* (Cambridge University Press, New York, 2001)
30. R. Schneider, *Convex Bodies: The Brunn-Minkowski Theory* (Cambridge University Press, New York, 1993)
31. A. Stancu, Centro-affine invariants for smooth convex bodies. Int. Math. Res. Not. **2012**, 2289–2320 (2012). https://doi.org/10.1093/imrn/rnr110
32. A. Stancu, Some affine invariants revisited, in *Asymptotic Geometric Analysis*, ed. by M. Ludwig et al. Fields Institute Communications (Springer, New York, 2013), pp. 343–359
33. E. Werner, D. Ye, New L_p affine isoperimetric inequalities. Adv. Math. **218**, 762–780 (2008)
34. E. Werner, D. Ye, Inequalities for mixed p-affine surface area. Math. Ann. **347**, 703–737 (2010)
35. Y. Yang, Moving frame and integrable system of discrete centroaffine curves in $\mathbb{R}^3$. Preprint (2016)

Characterizing the Volume via a Brunn-Minkowski Type Inequality

Jesús Yepes Nicolás

Abstract The Brunn-Minkowski inequality asserts that the n-th root of the functional volume is concave, namely, $\mathrm{vol}\big((1-\lambda)A + \lambda B\big)^{1/n}$ is greater than $(1-\lambda)\mathrm{vol}(A)^{1/n} + \lambda \mathrm{vol}(B)^{1/n}$ for compact sets A, B and $\lambda \in [0,1]$. Here we will show that if a given measure satisfies an inequality like this, with a certain positive power, for the family of all Euclidean balls then it must be a constant multiple of the volume.

1 Introduction

Let $\mathcal{K}^n$ be the set of all convex bodies, i.e., nonempty compact convex sets, in the n-dimensional Euclidean space $\mathbb{R}^n$, and let $|\cdot|$ denote the Euclidean norm in $\mathbb{R}^n$. We write B_n for the n-dimensional Euclidean (closed) unit ball whereas $\mathcal{B}^n$ will denote the set of all closed balls in $\mathbb{R}^n$, i.e.,

$$\mathcal{B}^n = \{x + rB_n : x \in \mathbb{R}^n, r > 0\}.$$

The volume of a measurable set $M \subset \mathbb{R}^n$, i.e., its n-dimensional Lebesgue measure, is denoted by $\mathrm{vol}(M)$ or $\mathrm{vol}_n(M)$ if the distinction of the dimension is useful (when integrating, as usual, $\mathrm{d}x$ will stand for $\mathrm{dvol}(x)$). With $\mathrm{int}\,M$, $\mathrm{bd}\,M$, $\mathrm{aff}\,M$ and $\dim M$ we represent its interior, boundary, affine hull and dimension (namely, the dimension of its affine hull), respectively.

Relating the volume with the Minkowski (vectorial) addition of convex bodies, one is led to the famous *Brunn-Minkowski inequality*. One form of it states that if $K, L \in \mathcal{K}^n$ and $\lambda \in (0,1)$, then

$$\mathrm{vol}\big((1-\lambda)K + \lambda L\big)^{1/n} \geq (1-\lambda)\mathrm{vol}(K)^{1/n} + \lambda \mathrm{vol}(L)^{1/n}, \tag{1.1}$$

J. Yepes Nicolás (✉)
Departamento de Matemáticas, Universidad de León, León, Spain
e-mail: jyepn@unileon.es

© Springer International Publishing AG 2018
G. Bianchi et al. (eds.), *Analytic Aspects of Convexity*, Springer INdAM Series 25,
https://doi.org/10.1007/978-3-319-71834-7_7

i.e., the n-th root of the volume is a concave function. Equality for some $\lambda \in (0, 1)$ holds if and only if K and L either lie in parallel hyperplanes or are homothetic.

The Brunn-Minkowski inequality is one of the most powerful theorems in Convex Geometry and beyond: it implies, among others, strong results such as the isoperimetric and Urysohn inequalities (see e.g. [19, s. 7.2]) or even the Aleksandrov-Fenchel inequality (see e.g. [19, s. 7.3]). It would not be possible to collect here all references regarding versions, applications and/or generalizations on the Brunn-Minkowski inequality. So, for extensive and beautiful surveys on them we refer the reader to [1, 7].

The Brunn-Minkowski inequality in its simplest form states that

$$\mathrm{vol}(K + L)^{1/n} \geq \mathrm{vol}(K)^{1/n} + \mathrm{vol}(L)^{1/n}, \tag{1.2}$$

from where one can verify its equivalence with (1.1) just by using the homogeneity of the volume. Yet some other equivalent forms of (1.1) are that

$$\mathrm{vol}\big((1 - \lambda)K + \lambda L\big) \geq \mathrm{vol}(K)^{1-\lambda}\mathrm{vol}(L)^{\lambda}, \tag{1.3}$$

which is often referred to in the literature as its multiplicative or dimension free form, and also that

$$\mathrm{vol}\big((1 - \lambda)K + \lambda L\big) \geq \min\{\mathrm{vol}(K), \mathrm{vol}(L)\}. \tag{1.4}$$

The main goal of this paper is to know whether the dimensional forms of the Brunn-Minkowski inequality, (1.1) and (1.2), are also satisfied for other measures on $\mathbb{R}^n$ or whether they constitute an inherent property of the volume. To this aim, we need first to overview some results closely related to this inequality.

Regarding an analytical counterpart for functions of the Brunn-Minkowski inequality, one is naturally led to the *Prékopa-Leindler inequality*, originally proved in [15] and [12].

Theorem A (The Prékopa-Leindler Inequality) *Let* $\lambda \in (0, 1)$ *and let* $f, g, h :$ $\mathbb{R}^n \longrightarrow \mathbb{R}_{\geq 0}$ *be non-negative measurable functions such that*

$$h\big((1 - \lambda)x + \lambda y\big) \geq f(x)^{1-\lambda}g(y)^{\lambda}$$

for all $x, y \in \mathbb{R}^n$. *Then*

$$\int_{\mathbb{R}^n} h(x)\, \mathrm{d}x \geq \left(\int_{\mathbb{R}^n} f(x)\, \mathrm{d}x\right)^{1-\lambda} \left(\int_{\mathbb{R}^n} g(x)\, \mathrm{d}x\right)^{\lambda}. \tag{1.5}$$

In fact, a straightforward proof of (1.3) can be obtained by applying (1.5) to the characteristic functions $f = \chi_K$, $g = \chi_L$ and $h = \chi_{(1-\lambda)K+\lambda L}$.

To further understand how the Prékopa-Leindler inequality is strongly related to the general Brunn-Minkowski inequality (1.1) one must know the so-called

Borell-Brascamp-Lieb inequality. In order to introduce it, we first recall the definition of the p-th mean of two non-negative numbers, where p is a parameter varying
in $\mathbb{R} \cup \{\pm\infty\}$ (for a general reference for p-means of non-negative numbers, we
refer the reader to the classic text of Hardy, Littlewood, and Pólya [9] and to the
handbook [4]). Consider first the case $p \in \mathbb{R}$ and $p \neq 0$; given $a, b \geq 0$ such that
$ab \neq 0$ and $\lambda \in (0, 1)$, we set

$$M_p(a, b, \lambda) = ((1 - \lambda)a^p + \lambda b^p)^{1/p}.$$

For $p = 0$ we define

$$M_0(a, b, \lambda) = a^{1-\lambda} b^\lambda$$

and, to complete the picture, for $p = \pm\infty$ we set $M_{+\infty}(a, b, \lambda) = \max\{a, b\}$ and
$M_{-\infty}(a, b, \lambda) = \min\{a, b\}$. Finally, if $ab = 0$, we will define $M_p(a, b, \lambda) = 0$ for
all $p \in \mathbb{R} \cup \{\pm\infty\}$. Note that $M_p(a, b, \lambda) = 0$, if $ab = 0$, is redundant for all $p \leq 0$,
however it is relevant for $p > 0$. Furthermore, for $p \neq 0$, we shall allow that a, b
take the value $+\infty$ and in that case, as usual, $M_p(a, b, \lambda)$ will be the value that is
obtained "by continuity" with respect to p.

Jensen's inequality for means (see e.g [9, Section 2.9] and [4, Theorem 1 p. 203])
implies that if $-\infty \leq p < q < +\infty$ then

$$M_p(a, b, \lambda) \leq M_q(a, b, \lambda), \tag{1.6}$$

with equality for $ab > 0$ and $\lambda \in (0, 1)$ if and only if $a = b$.

The following theorem contains the Borell-Brascamp-Lieb inequality (see [2, 3]
and also [7] for a detailed presentation), which uses the p-th mean to generalize the
Prékopa-Leindler inequality (the case $p = 0$).

Theorem B (The Borell-Brascamp-Lieb Inequality) *Let $\lambda \in (0, 1)$, $-1/n \leq q \leq$
$+\infty$ and let $f, g, h : \mathbb{R}^n \longrightarrow \mathbb{R}_{\geq 0}$ be non-negative measurable functions such that*

$$h\big((1 - \lambda)x + \lambda y\big) \geq M_q \left(f(x), g(y), \lambda \right)$$

for all $x, y \in \mathbb{R}^n$. Then

$$\int_{\mathbb{R}^n} h(x) \, dx \geq M_p \left(\int_{\mathbb{R}^n} f(x) \, dx, \int_{\mathbb{R}^n} g(x) \, dx, \lambda \right),$$

where $p = q/(nq + 1)$.

As a direct application of this result we notice that, for $f = \chi_K$, $g = \chi_L$ and $h =$
$\chi_{(1-\lambda)K + \lambda L}$, (1.1) is obtained when taking $q = +\infty$ whereas (1.4) holds if we set
$q = -1/n$ (and thus $p = -\infty$).

Regarding the functions which are naturally connected to the above theorem, we
get to the following definition (see e.g. [3]).

Definition 1.1 A non-negative function $f : \mathbb{R}^n \longrightarrow \mathbb{R}_{\geq 0}$ is q-concave, for a given $q \in \mathbb{R} \cup \{\pm\infty\}$, if for all $x, y \in \mathbb{R}^n$ and all $\lambda \in (0, 1)$,

$$f\big((1 - \lambda)x + \lambda y\big) \geq M_q \left(f(x), f(y), \lambda\right).$$

A 0-concave function is usually called *log-concave* whereas a $(-\infty)$-concave function is referred to as *quasi-concave*.

Let μ be a measure on $\mathbb{R}^n$ with density function $f : \mathbb{R}^n \longrightarrow \mathbb{R}_{\geq 0}$. If f is q-concave, with $-1/n \leq q \leq +\infty$, then by the Borell-Brascamp-Lieb inequality for $\overline{f} = f \chi_A$, $\overline{g} = f \chi_B$ and $\overline{h} = f \chi_{(1-\lambda)A+\lambda B}$, we have that

$$\mu((1 - \lambda)A + \lambda B) \geq M_p(\mu(A), \mu(B), \lambda) = \big((1 - \lambda)\mu(A)^p + \lambda \mu(B)^p\big)^{1/p} \quad (1.7)$$

for any pair of measurable sets A, B with $\mu(A)\mu(B) > 0$ and such that $(1 - \lambda)A + \lambda B$ is also measurable, where $p = q/(nq + 1)$. From now on, a measure μ satisfying (1.7) will be said to be *p-concave*.

Borell [2, Theorem 3.2] (see also [6, Section 3.3]) gave a sort of converse to this statement:

Theorem C *Let $-\infty \leq p \leq 1/n$ and let μ be a Radon measure on an open convex set $\Omega \subset \mathbb{R}^n$, which is also its support. If μ is p-concave on Ω then there exists a q-concave function f such that $d\mu(x) = f(x)dx$, where $-1/n \leq q \leq +\infty$ is so that $p = q/(nq + 1)$.*

In other words, p-concave measures are associated to q-concave functions (under the link $p = q/(nq + 1)$) and vice versa. When $0 < p < 1/n$, q is non-negative and then f is q-concave if and only if f^q is concave on the convex set $\Omega = \{x \in \mathbb{R}^n : f(x) > 0\}$. Thus, on the one hand, since there are no further non-negative concave functions defined on the whole Euclidean space $\mathbb{R}^n$ than constants (see e.g. [16, Problem-Remark H, p. 8]), if μ is a Radon measure with support $\mathbb{R}^n$ that is p-concave, for some $p > 0$, then it must be, up to a constant, the volume. On the other hand, (for $p = 1/n$) since the sole $(+\infty)$-concave functions supported on open sets are those that are constants over them (see e.g. [3, p. 373]), if μ is a Radon measure supported on a certain open convex subset of $\mathbb{R}^n$ that is $(1/n)$-concave then the only possibility, once again, is that it is a constant multiple of the volume.

Here we are interested in showing this characterization of the volume, via the Brunn-Minkowski inequality, independently of Borell's result. Moreover, we will prove, on the one hand, that it is enough to assume that the measure satisfies the corresponding Brunn-Minkowski inequality for a subfamily of convex bodies: the set of all balls $\mathcal{B}^n$. More precisely, the main result of the paper reads as follows.

Theorem 1.2 *Let $p > 0$ and let $\Omega \subset \mathbb{R}^n$ be an open convex set. Let μ be a locally finite Borel measure on Ω such that*

$$\mu((1 - \lambda)K + \lambda L)^p \geq (1 - \lambda)\mu(K)^p + \lambda \mu(L)^p \quad (1.8)$$

holds for any pair of balls $K, L \in \mathcal{B}^n$ *with* $K, L \subset \Omega$, *and all* $\lambda \in (0, 1)$. *Then* $\mu = c \operatorname{vol}_n$ *for some (constant)* $c \geq 0$ *if either* $\Omega = \mathbb{R}^n$ *or* $p = 1/n$.

When dealing with Brunn-Minkowski type inequalities (cf. (1.7)), it is natural to wonder about the improvement of the concavity of the corresponding measure (see e.g. [8, 10, 13] and [14]), i.e., whether it is possible to 'enhance' the exponent p for such an inequality, in the sense of considering a tighter p-th mean (cf. (1.6)). To this aim, many times one shows that the desired inequality is not true for arbitrary convex bodies and thus, it is necessary to consider the problem only for special subfamilies of sets. In this context, the characterization provided in this paper can be viewed as a useful tool for this type of problems, as well as another step for a better understanding of the extent and diversity of the Brunn-Minkowski inequality and its applications.

We will also explicitly show that the only condition needed for the measure is being locally finite (see Lemma 2.3), as well as that both the assumptions $p > 0$ when the support of the measure is $\mathbb{R}^n$ and $p = 1/n$ when it is an arbitrary open convex set are necessary (see Remark 3.5).

In contrast to Theorem 1.2, the additive version of the Brunn-Minkowski inequality (cf. (1.2)) characterizes the volume even in the case in which neither the measure satisfies this inequality on the whole space $\mathbb{R}^n$ nor the exponent p equals $1/n$. The only assumption needed to this aim is assuming that the origin is an interior point. More precisely, we show the following result.

Theorem 1.3 *Let* $p > 0$ *and let* $\Omega \subset \mathbb{R}^n$ *be an open convex set with* $0 \in \Omega$. *Let* μ *be a locally finite Borel measure on* Ω *such that*

$$\mu(K + L)^p \geq \mu(K)^p + \mu(L)^p$$

holds for any pair of balls $K, L \in \mathcal{B}^n$ *with* $K, L \subset \Omega$, *and all* $\lambda \in (0, 1)$. *Then* $\mu = c \operatorname{vol}_n$ *for some (constant)* $c \geq 0$.

The paper is organized as follows. Section 2 is mainly devoted to collecting some definitions and auxiliary well-known results, whereas Sects. 3 and 4 are devoted to proving (among other results) Theorems 1.2 and 1.3. In Sect. 3 we will focus on the simpler case in which the measure is absolutely continuous with a continuous density function, for the purpose of showing the general case of an arbitrary locally finite measure along Sect. 4.

2 Background Material and Auxiliary Results

For the sake of completeness, we will collect some definitions and well-known facts from measure theory that will be used throughout this work. We refer the reader to [5].

Definition 2.1 Let $\Omega \subset \mathbb{R}^n$ be a nonempty set. Let (Ω, μ) be a measure space where μ is a Borel measure on Ω, i.e., we will omit for simplicity the σ-algebra Σ which will be assumed to contain the σ-algebra of Borel sets in Ω, and that will be contained in the σ-algebra of Lebesgue measurable sets.

We recall the definition of the support of a measure μ, which is the set $\operatorname{supp}(\mu) = \{x \in \Omega : \mu((x + rB_n) \cap \Omega) > 0 \text{ for all } r > 0\}$. When $\Omega = \mathbb{R}^n$, the support is closed because its complement is the union of the open sets of measure 0. Moreover, by compactness arguments, and writing $\mathbb{R}^n \setminus \operatorname{supp}(\mu)$ as a countable union of compact sets (cf. [5, Proposition 1.1.6]), we clearly have $\mu(\mathbb{R}^n \setminus \operatorname{supp}(\mu)) = 0$.

Definition 2.2 A measure μ on $\Omega \subset \mathbb{R}^n$ is said to be *locally finite* if for every point $x \in \Omega$ there exists $r = r(x) > 0$ such that $\mu(x + rB_n) < +\infty$. In particular, a locally finite measure is finite on every compact subset of Ω and hence is σ-*finite*, i.e., Ω is the countable union of measurable sets with finite measure.

Moreover, a measure μ is called a Radon measure if it is locally finite and $\mu(A) = \sup\{\mu(K) : K \subset A \text{ compact}\}$ for every measurable set A.

In this paper we deal with the characterization of measures on (subsets of) $\mathbb{R}^n$ via the Brunn-Minkowski inequality. To this end, from now on, we will omit the measures that are not locally finite because in that case they are trivially defined on the sets with nonempty interior. This is the content of the following result.

Lemma 2.3 *Let $p > 0$ and let $\Omega \subset \mathbb{R}^n$ be an open convex set. Let μ be a Borel measure on Ω that is not locally finite. If*

$$\mu((1 - \lambda)K + \lambda L)^p \geq (1 - \lambda)\mu(K)^p + \lambda\mu(L)^p \tag{2.1}$$

holds for any pair of balls $K, L \in \mathcal{B}^n$ with $K, L \subset \Omega$, and all $\lambda \in (0, 1)$, then $\mu(A) = +\infty$ for all $A \subset \Omega$ with nonempty interior.

Proof It is enough to show that $\mu(B) = +\infty$ for all $B \in \mathcal{B}^n$, $B \subset \Omega$. Since μ is not locally finite, there exists $x_0 \in \Omega$ such that

$$\mu(x_0 + rB_n) = +\infty$$

for all $r > 0$ such that $x_0 + rB_n \subset \Omega$. Since Ω is open, for any $x \in \Omega$, $x \neq x_0$, there exist $y \in \Omega$ and $\lambda \in (0, 1)$ such that $(1 - \lambda)x_0 + \lambda y = x$. Thus, taking $K = x_0 + rB_n$ and $L = y + rB_n$ for $r > 0$ small enough, by (2.1) we have

$$\mu(x + rB_n)^p \geq (1 - \lambda)\mu(x_0 + rB_n)^p + \lambda\mu(y + rB_n)^p = +\infty, \tag{2.2}$$

which concludes the proof. $\square$

We would like to point out that the proof of the precedent result does not work for further values of p, because in that case we would also have (cf. (2.2)) that $\mu(x + rB_n) \geq M_p(\mu(x_0 + rB_n), \mu(y + rB_n), \lambda)$, but the right-hand side would not be, in general, infinity.

Definition 2.4 We recall that μ is said to be *concentrated* on a measurable set $A \subset \Omega$ if $\mu(\Omega \setminus A) = 0$. In this sense, μ is *singular* with respect to the Lebesgue measure if there exists a measurable set $A \subset \Omega$ such that μ is concentrated on A and vol is concentrated on $\Omega \setminus A$.

Conversely, μ is said to be *absolutely continuous* (with respect to the Lebesgue measure) if for every measurable set $A \subset \Omega$, $\mu(A) = 0$ whenever vol$(A) = 0$.

We would like to stress that, for simplicity, we will sometimes omit the Lebesgue measure in those notions, like absolutely continuous or singular, which refer to a specific measure in a context where more than one are mentioned. In the same way, expressions like *for almost every x* will mean for vol-*almost every x*.

The well-known Radon-Nikodym theorem (see e.g. [5, Theorem 4.2.2]) will play a relevant role throughout this paper.

Theorem D (Radon-Nikodym Theorem) *Let μ be a σ-finite measure on $\Omega \subset \mathbb{R}^n$. If μ is absolutely continuous (with respect to the Lebesgue measure) then there exists a measurable function $f : \Omega \longrightarrow \mathbb{R}_{\geq 0}$ such that*

$$\mu(A) = \int_A f(x)\, dx$$

for any measurable set $A \subset \Omega$.

Such a function f for a given σ-finite absolutely continuous measure is usually called a density function of μ, which will be denoted, for short, as $d\mu(x) = f(x)dx$.

Another useful tool along the paper will be the Lebesgue decomposition theorem (we refer the reader to [5, Theorem 4.3.2]), which asserts that, roughly speaking, every σ-finite measure is the sum of an absolutely continuous measure and a singular one.

Theorem E (Lebesgue's Decomposition Theorem) *Let μ be a σ-finite Borel measure on $\Omega \subset \mathbb{R}^n$. Then there is a unique measure μ_s and a unique (up to a null set) measurable function $f : \Omega \longrightarrow \mathbb{R}_{\geq 0}$ such that*

$$d\mu(x) = f(x)dx + d\mu_s(x), \tag{2.3}$$

where μ_s is singular with respect to the Lebesgue measure.

For a σ-finite Borel measure μ, (2.3) will be referred to as the Lebesgue decomposition of μ (with respect to the Lebesgue measure).

The following result (see e.g. [18, Theorem 7.7]) shows that the content of the fundamental theorem of calculus over the real line persists in the setting of the Lebesgue integral over the whole Euclidean space $\mathbb{R}^n$. To this end one must consider the so-called symmetric derivative of the measure μ given by $d\mu(x) = f(x)dx$. First we recall the following definition (see e.g. [5]).

Definition 2.5 A function $f : \Omega \subset \mathbb{R}^n \longrightarrow \mathbb{R}_{\geq 0}$ is *locally integrable* if for every point $x \in \Omega$ there exists $r = r(x) > 0$ such that $\int_{x+rB_n} f(t)\,dt < +\infty$. In particular, the integral of a locally integrable function is finite on every compact subset of Ω.

Theorem F (Lebesgue's Differentiation Theorem) *Let $\Omega \subset \mathbb{R}^n$ be an open set and let $f : \Omega \longrightarrow \mathbb{R}_{\geq 0}$ be a locally integrable function. Then*

$$\lim_{r\to 0^+} \frac{1}{\mathrm{vol}(rB_n)} \int_{x+rB_n} f(t)\,dt = f(x)$$

for almost every $x \in \Omega$.

Furthermore, if f is a continuous function then the above condition holds everywhere.

The above result admits a stronger version in the setting of the Radon-Nikodym theorem and the Lebesgue decomposition, as the following result shows (see e.g. [18, Theorem 7.14]).

Theorem G *Let $\Omega \subset \mathbb{R}^n$ be an open set, let μ be a locally finite Borel measure on Ω, and let $d\mu(x) = f(x)dx + d\mu_s(x)$ be the Lebesgue decomposition of μ. Then*

$$\lim_{r\to 0^+} \frac{\mu(x+rB_n)}{\mathrm{vol}(rB_n)} = f(x)$$

for almost every $x \in \Omega$.

When working with singular measures, the symmetric derivative satisfies the following property (see e.g. [18, Theorem 7.15]).

Theorem H *Let $\Omega \subset \mathbb{R}^n$ be an open set and let μ be a Borel measure on Ω that is singular with respect to the Lebesgue measure. Then*

$$\lim_{r\to 0^+} \frac{\mu(x+rB_n)}{\mathrm{vol}(rB_n)} = +\infty$$

for μ-almost every $x \in \Omega$.

By Definition 2.4, we notice that Theorems G and H suppose to be the two faces of the same coin, in the sense that each of them shows what essentially happens in the sets where the absolutely continuous part and the singular one of a given locally finite measure are, respectively, concentrated.

3 Simple Case: Absolutely Continuous Measures with Continuous Radon-Nikodym Derivative

Here we will show the statement of Theorem 1.2 when working with absolutely continuous measures associated to continuous density functions. In other words, we will prove on the one hand that, for such a measure, assuming the Brunn-Minkowski inequality with exponent $p > 0$ (cf. (1.8)) in the whole Euclidean space $\mathbb{R}^n$, is equivalent to say that the measure is (up to a constant) the volume. On the other hand, the Brunn-Minkowski inequality with exponent $1/n$ in a given open convex set $\Omega \subset \mathbb{R}^n$ yields the same consequence. Moreover, the assumptions that either $\Omega = \mathbb{R}^n$ for a given $p > 0$ or $p = 1/n$ for an arbitrary open convex set $\Omega \subset \mathbb{R}^n$, are necessary (see Remark 3.5).

Furthermore, we would like to point out that along this paper we will not assume, in principle, that the exponent p is not bigger than $1/n$. However, we will get this constraint for p (unless we are dealing with the zero measure). Indeed, if the measure satisfies the Brunn-Minkowski inequality for $p > 1/n$, then it also does for $p = 1/n$ because of (1.6) and thus, from Theorem 1.2, the measure must be a constant multiple of the volume. The equality case of (1.1) (together with (1.6)) yields that $1/n$ is the 'largest' exponent for such an inequality for the volume, and hence this implies that the given measure must be the zero one.

We will start by showing that, if an absolutely continuous measure, with continuous Radon-Nikodym derivative, satisfies the Brunn-Minkowski inequality with exponent p then its density function must be quasi-concave (even when $p \leq 0$).

Lemma 3.1 *Let $p \in \mathbb{R} \cup \{\pm\infty\}$ and let μ be the measure on $\mathbb{R}^n$ given by $d\mu(x) = f(x)dx$, where f is a (non-negative) continuous function. If*

$$\mu((1-\lambda)K + \lambda L) \geq \left((1-\lambda)\mu(K)^p + \lambda\mu(L)^p\right)^{1/p}$$

holds for any pair of balls $K, L \in \mathcal{B}^n$, and all $\lambda \in (0,1)$, then f is quasi-concave.

Proof Suppose, by contradiction, that $f((1-\lambda_0)x + \lambda_0 y) < \min\{f(x), f(y)\}$ for certain $x, y \in \mathbb{R}^n$ and $\lambda_0 \in (0,1)$. Let $z = (1-\lambda_0)x + \lambda_0 y$ and let $\varepsilon > 0$ be such that $f(z) + \varepsilon < \min\{f(x), f(y)\} - \varepsilon$.

Since f is continuous there exists $\delta > 0$ such that $|f(t) - f(t')| < \varepsilon$ for all $t \in t' + \delta B_n$ and $t' \in \{x, y, z\}$. Then, taking $K = x + \delta B_n$ and $L = y + \delta B_n$, we have

$$\mu((1-\lambda_0)K + \lambda_0 L) = \mu(z + \delta B_n) = \int_{z+\delta B_n} f(t)\, dt$$

$$\leq \delta^n \kappa_n (f(z) + \varepsilon) < \delta^n \kappa_n \left(\min\{f(x), f(y)\} - \varepsilon\right)$$

$$\leq \left[(1-\lambda_0)\left(\int_{x+\delta B_n} f(t)\, dt\right)^p + \lambda_0 \left(\int_{y+\delta B_n} f(t)\, dt\right)^p\right]^{1/p}$$

$$= \left((1-\lambda_0)\mu(K)^p + \lambda_0\mu(L)^p\right)^{1/p},$$

a contradiction.   □

Next we will show the main result of this section, which is the particular case of Theorem 1.2 for absolutely continuous measures with continuous density function and $\Omega = \mathbb{R}^n$.

Theorem 3.2 *Let $p > 0$ and let μ be the measure on $\mathbb{R}^n$ given by $d\mu(x) = f(x)dx$, where f is a (non-negative) continuous function. If*

$$\mu((1 - \lambda)K + \lambda L)^p \geq (1 - \lambda)\mu(K)^p + \lambda\mu(L)^p$$

holds for any pair of balls $K, L \in \mathcal{B}^n$, and all $\lambda \in (0, 1)$, then $\mu = c \operatorname{vol}_n$ for some (constant) $c \geq 0$.

Proof Suppose, by contradiction, that $f(x) \neq f(y)$ for some $x, y \in \mathbb{R}^n$. Without loss of generality, $\operatorname{aff}(\{x, y\}) = \{z \in \mathbb{R}^n : z_2 = \cdots = z_n = 0\}$. Let $f_1 : \mathbb{R} \longrightarrow \mathbb{R}_{\geq 0}$ be the function given by $f_1(s) = f(s, 0, \ldots, 0)$. Then $x = (x_1, 0, \ldots, 0)$, $y = (y_1, 0 \ldots, 0)$ and $f(x) = f_1(x_1), f(y) = f_1(y_1)$.

Without loss of generality, we may assume that $f_1(x_1) > f_1(y_1)$. By Lemma 3.1, f is quasi-concave and then f_1 is so. Hence, assuming that $x_1 < y_1$, the quasi-concavity of f_1 implies that it must be decreasing in $[y_1, +\infty)$ and thus, the limit $0 \leq L = \lim_{t \to +\infty} f_1(t)$ exists (the case $x_1 > y_1$ is analogous). So, $L \leq f_1(y_1) < f_1(x_1)$ and thus we can find $\varepsilon > 0$ and $\lambda_0 \in (0, 1)$ such that

$$L + \varepsilon < (f_1(x_1) - \varepsilon)(1 - \lambda_0)^{1/p}. \tag{3.1}$$

Moreover, for such an ε, there exists $z_1 \in \mathbb{R}$ such that

$$f_1(t) \leq L + \frac{\varepsilon}{2} \tag{3.2}$$

for all $t \geq z_1$.

On the other hand, since f is continuous, there exists $\delta > 0$ such that

$$\left| f(t) - f(t') \right| < \frac{\varepsilon}{2} \tag{3.3}$$

for all $t \in t' + \delta B_n$ and $t' \in \{x, z\}$, where $z = (z_1, 0 \ldots, 0)$.

Let $K = x + \delta B_n$ and let $L = w + \delta B_n$, where $w = (w_1, 0 \ldots, 0)$ is so that $(1 - \lambda_0)x + \lambda_0 w = z$. Then, by (3.1)–(3.3), we have

$$\mu((1 - \lambda_0)K + \lambda_0 L) = \mu(z + \delta B_n) = \int_{z + \delta B_n} f(t)\, dt$$

$$\leq \delta^n \kappa_n \left(f(z) + \frac{\varepsilon}{2} \right) = \delta^n \kappa_n \left(f_1(z_1) + \frac{\varepsilon}{2} \right) \leq \delta^n \kappa_n (L + \varepsilon)$$

$$< \delta^n \kappa_n (f_1(x_1) - \varepsilon)(1 - \lambda_0)^{1/p} = \delta^n \kappa_n (f(x) - \varepsilon)(1 - \lambda_0)^{1/p}$$

$$\leq \left[(1 - \lambda_0) \left(\int_{x + \delta B_n} f(t)\, dt \right)^p \right]^{1/p} = \left((1 - \lambda_0)\mu(K)^p \right)^{1/p}$$

$$\leq \left((1 - \lambda_0)\mu(K)^p + \lambda_0\mu(L)^p \right)^{1/p},$$

a contradiction. $\square$

Second Proof of Theorem 3.2 For any $x, y \in \mathbb{R}^n$ and $\lambda \in [0, 1]$, we have

$$\left(\int_{(1-\lambda)x+\lambda y+rB_n} f(t)\, dt \right)^p \geq (1 - \lambda) \left(\int_{x+rB_n} f(t)\, dt \right)^p + \lambda \left(\int_{y+rB_n} f(t)\, dt \right)^p$$

for all $r > 0$ and thus, dividing by $\mathrm{vol}(rB_n)^p$ and using Theorem F, we get

$$f((1 - \lambda)x + \lambda y)^p \geq (1 - \lambda)f(x)^p + \lambda f(y)^p.$$

Then f^p is a concave function on the whole $\mathbb{R}^n$ and thus, since f is non-negative, f must be constant (see e.g. [16, Problem-Remark H, p. 8]).   □

Now, we deal with the particular case $p = 1/n$ of Theorem 1.2, i.e., assuming the Brunn-Minkowski inequality with exponent $1/n$ over an arbitrary open convex set Ω. To this aim, we will start with the one-dimensional case, since the proof we present here will help us to better understand which approach 'should' be carried out for the corresponding n-dimensional case, shown in Theorem 3.4.

Theorem 3.3 *Let $\Omega \subset \mathbb{R}$ be an open convex set and let μ be the measure on Ω given by $d\mu(x) = f(x)dx$, where f is a (non-negative) continuous function. If*

$$\mu((1 - \lambda)K + \lambda L) \geq (1 - \lambda)\mu(K) + \lambda\mu(L) \tag{3.4}$$

holds for any pair of balls $K, L \in \mathcal{B}^1$ with $K, L \subset \Omega$, and all $\lambda \in (0, 1)$, then $\mu = c\, \mathrm{vol}_1$ for some (constant) $c \geq 0$.

Proof Let $x_0 \in \Omega$ and let $F : \Omega \longrightarrow \mathbb{R}$ be the function given by

$$F(x) = \int_{x_0}^x f(t)\, dt.$$

Fix $x, y > x_0$, $x, y \in \Omega$, and take $K = [x_0, x]$ and $L = [x_0, y]$. Then, from (3.4), we get $F\big((1 - \lambda)x + \lambda y\big) \geq (1 - \lambda)F(x) + \lambda F(y)$. Since it is true for arbitrary $x, y \in (x_0, +\infty) \cap \Omega$ and $\lambda \in [0, 1]$, we may assure that F is concave on $(x_0, +\infty) \cap \Omega$. In the same way, we obtain that F is convex on $(-\infty, x_0) \cap \Omega$.

Moreover, since f is continuous, by the fundamental theorem of calculus we get $F'(x) = f(x)$, for all $x \in \Omega$. Now, the concavity of F on $(x_0, +\infty) \cap \Omega$ (resp. the convexity of F on $(-\infty, x_0) \cap \Omega$) implies that $f(x) = F'(x)$ is decreasing in $(x_0, +\infty) \cap \Omega$ (resp. $f(x) = F'(x)$ is increasing in $(-\infty, x_0) \cap \Omega$). Since $x_0 \in \Omega$ is arbitrary, f must be constant.   □

As we have just seen in the above result, the 'local nature' of the Brunn-Minkowski inequality on (a subset of) $\mathbb{R}$ suggests us to employ some tools from differential calculus such as its fundamental theorem. Thus, for the general case, it seems to be natural to use the n-dimensional counterpart of the above-mentioned result in the setting of the Lebesgue integral, i.e., the Lebesgue differentiation theorem (Theorem F).

Moreover, we would like to point out that, although Borell's approach in [2] is quite involved and complicated, the underlying key idea of his proof of Theorem C is exploiting the Lebesgue differentiation theorem for boxes with suitable lengths (depending on the range in which the parameter p lies) in order to get the corresponding desired properties of concavity of the density function. This idea is similar to that of the proof we present here for the following result, which is included for the sake of completeness.

Theorem 3.4 *Let $\Omega \subset \mathbb{R}^n$ be an open convex set and let μ be the measure on Ω given by $d\mu(x) = f(x)dx$, where f is a (non-negative) continuous function. If*

$$\mu((1-\lambda)K + \lambda L)^{1/n} \geq (1-\lambda)\mu(K)^{1/n} + \lambda\mu(L)^{1/n} \tag{3.5}$$

holds for any pair of balls $K, L \in \mathcal{B}^n$ with $K, L \subset \Omega$, and all $\lambda \in (0,1)$, then $\mu = c\,\mathrm{vol}_n$ for some (constant) $c \geq 0$.

Proof Let $r_1, r_2 > 0$ be fixed. Then, for any $x, y \in \Omega$ and $\lambda \in [0,1]$, by (3.5) we have

$$\left(\int_{(1-\lambda)x+\lambda y + r((1-\lambda)r_1+\lambda r_2)B_n} f(t)\,dt\right)^{1/n}$$

$$\geq (1-\lambda)\left(\int_{x+rr_1B_n} f(t)\,dt\right)^{1/n} + \lambda\left(\int_{y+rr_2B_n} f(t)\,dt\right)^{1/n}$$

for all $r > 0$ small enough. Thus, dividing by $\mathrm{vol}(rB_n)^{1/n}$ and using Theorem F (together with the fact that the volume is homogeneous of degree n), we get

$$f((1-\lambda)x + \lambda y)^{1/n}((1-\lambda)r_1 + \lambda r_2) \geq (1-\lambda)f(x)^{1/n}r_1 + \lambda f(y)^{1/n}r_2.$$

Now, by taking limits in the latter expression as $r_2 \to 0$ and $r_1 \to 0$, respectively, we may assure that $f((1-\lambda)x + \lambda y) \geq \max\{f(x), f(y)\}$ (for any $x, y \in \Omega$ and all $\lambda \in [0,1]$). Hence, and since Ω is open, f must be constant. □

Remark 3.5 In relation to Theorem 1.2, we would like to point out that the assumptions $p > 0$ when $\Omega = \mathbb{R}^n$ or $p = 1/n$ when Ω is an arbitrary open convex set, are necessary.

Indeed, for $p = 0$ (and hence, from the monotonicity of the p-means (1.6), also for any $p < 0$) it is enough to consider the standard Gaussian measure γ in $\mathbb{R}^n$, which is given by

$$d\gamma(x) = \frac{1}{(2\pi)^{n/2}}e^{\frac{-|x|^2}{2}}\,dx,$$

because it is log-concave and thus, by Theorem A, it satisfies the (multiplicative) Brunn-Minkowski inequality on the whole $\mathbb{R}^n$, i.e., (1.7) holds for any pair of convex bodies $K, L \in \mathcal{K}^n$ and all $\lambda \in (0, 1)$.

On the other hand, let $q \in \mathbb{R}_{>0}$ and let μ_q be the measure given by

$$d\mu_q(x) = (1 - |x|)^{1/q} \chi_{B_n}(x) \, dx.$$

Then, by the Borell-Brascamp-Lieb inequality, Theorem B, μ_q satisfies the Brunn-Minkowski inequality (1.7) for $p = q/(nq + 1)$. Thus, taking $\Omega = \text{int } B_n$, (1.8) holds for any pair of non-degenerate convex bodies $K, L \in \mathcal{K}^n$ with $K, L \subset \Omega$, and all $\lambda \in (0, 1)$.

4 General Case

The main goal of this section is to show Theorems 1.2 and 1.3. To this aim, we will prove that it is enough to work with absolutely continuous measures with continuous density function, and thus we may use the results that were previously obtained in Sect. 3. More precisely, we will show the following:

Lemma 4.1 *Let $p > 0$ and let $\Omega \subset \mathbb{R}^n$ be an open convex set. Let μ be a locally finite Borel measure on Ω such that*

$$\mu((1 - \lambda)K + \lambda L)^p \geq (1 - \lambda)\mu(K)^p + \lambda\mu(L)^p$$

holds for any pair of balls $K, L \in \mathcal{B}^n$ with $K, L \subset \Omega$, and all $\lambda \in (0, 1)$. Then μ is absolutely continuous and $d\mu(x) = f(x)dx$ where $f : \Omega \longrightarrow \mathbb{R}_{\geq 0}$ is continuous.

For the sake of simplicity, we will split the above result into another two, namely, Lemmas 4.2 and 4.4. We will start this section by showing that a locally finite measure that satisfies the Brunn-Minkowski inequality (1.8) is absolutely continuous. We would like to stress here the relevance of the assumption of locally finiteness, in contrast to what Lemma 2.3 ensures.

Lemma 4.2 *Let $p > 0$ and let $\Omega \subset \mathbb{R}^n$ be an open convex set. Let μ be a locally finite Borel measure on Ω such that*

$$\mu((1 - \lambda)K + \lambda L)^p \geq (1 - \lambda)\mu(K)^p + \lambda\mu(L)^p \tag{4.1}$$

holds for any pair of balls $K, L \in \mathcal{B}^n$ with $K, L \subset \Omega$, and all $\lambda \in (0, 1)$. Then μ is absolutely continuous.

Proof Since μ is locally finite, by Theorem E, there exist a singular measure μ_s and a measurable function $f : \Omega \longrightarrow \mathbb{R}_{\geq 0}$ for which $d\mu(x) = f(x)dx + d\mu_s(x)$.

Moreover, by means of Theorem H, the set

$$A = \left\{ x \in \Omega : \lim_{r \to 0^+} \frac{\mu_s(x + rB_n)}{\mathrm{vol}(rB_n)} = +\infty \right\}$$

satisfies that $\mu_s(\Omega \setminus A) = 0$.

If A is nonempty, there exists $x_0 \in \Omega$ for which

$$\lim_{r \to 0^+} \frac{\mu_s(x_0 + rB_n)}{\mathrm{vol}(rB_n)} = +\infty. \qquad (4.2)$$

Since Ω is open, for any $x \in \Omega$, $x \neq x_0$, there exist $y \in \Omega$ and $\lambda \in (0,1)$ such that $(1-\lambda)x_0 + \lambda y = x$. Then, by (4.1), for all $r > 0$ small enough we get

$$\mu(x + rB_n) \geq \left((1-\lambda)\mu(x_0 + rB_n)^p + \lambda\mu(y + rB_n)^p \right)^{1/p}$$

$$\geq (1-\lambda)^{1/p}\mu(x_0 + rB_n) \geq (1-\lambda)^{1/p}\mu_s(x_0 + rB_n).$$

Now, the above inequality implies, by (4.2), that

$$\lim_{r \to 0^+} \frac{\mu(x + rB_n)}{\mathrm{vol}(rB_n)} = +\infty$$

for all $x \in \Omega$, a contradiction with the statement of Theorem G (we notice that $f(x) \neq +\infty$ for all $x \in \Omega$). Thus, A is empty and hence μ_s is identically zero. □

Remark 4.3 We would like to stress, for the sake of completeness, the role played by both Theorem G and Theorem H in the proof of the precedent result, Lemma 4.2.

On the one hand, regarding the singular part of μ, μ_s, we have that the set

$$A = \left\{ x \in \Omega : \lim_{r \to 0^+} \frac{\mu_s(x + rB_n)}{\mathrm{vol}(rB_n)} = +\infty \right\}$$

satisfies that $\mu_s(\Omega \setminus A) = 0$, because of Theorem H (and moreover, $\mathrm{vol}(A) = 0$, by Theorem G).

On the other hand, Theorem G together with the p-concavity of μ imply, using the above consequence of Theorem H, that A must be empty and thus μ_s is identically zero.

In order that the general case can be reduced to the one studied in the previous section, we must show that the Radon-Nikodym derivative can be chosen to be continuous. This is the content of the following result, which is proved with a quite standard argument, and whose proof is included for the sake of completeness.

Lemma 4.4 *Let $p > 0$ and let $\Omega \subset \mathbb{R}^n$ be an open convex set. Let μ be a locally finite Borel measure on Ω given by $\mathrm{d}\mu(x) = f(x)\mathrm{d}x$. If*

$$\mu((1 - \lambda)K + \lambda L)^p \geq (1 - \lambda)\mu(K)^p + \lambda\mu(L)^p$$

holds for any pair of balls $K, L \in \mathcal{B}^n$ with $K, L \subset \Omega$, and all $\lambda \in (0, 1)$, then there exists a (non-negative) continuous function $\phi : \Omega \longrightarrow \mathbb{R}_{\geq 0}$ such that $\phi(x) = f(x)$ for almost every $x \in \Omega$.

Proof Let $\phi : \Omega \longrightarrow \mathbb{R}_{\geq 0} \cup \{+\infty\}$ be the function given by

$$\phi(x) = \liminf_{r \to 0^+} \frac{1}{\mathrm{vol}(rB_n)} \int_{x + rB_n} f(t)\, \mathrm{d}t.$$

On the one hand, we have that $\phi(x) = f(x)$ for almost every $x \in \Omega$, by Theorem F (we notice that f is locally integrable because μ is locally finite).

On the other hand, for any $x, y \in \Omega$ and $\lambda \in [0, 1]$,

$$\phi((1 - \lambda)x + \lambda y) = \liminf_{r \to 0^+} \frac{1}{\mathrm{vol}(rB_n)} \int_{(1-\lambda)x + \lambda y + rB_n} f(t)\, \mathrm{d}t$$

$$\geq \liminf_{r \to 0^+} \frac{1}{\mathrm{vol}(rB_n)} \left[(1 - \lambda) \left(\int_{x + rB_n} f(t)\, \mathrm{d}t \right)^p + \lambda \left(\int_{y + rB_n} f(t)\, \mathrm{d}t \right)^p \right]^{1/p}$$

$$\geq \left[(1 - \lambda) \left(\liminf_{r \to 0^+} \frac{1}{\mathrm{vol}(rB_n)} \int_{x + rB_n} f(t)\, \mathrm{d}t \right)^p \right.$$

$$+ \lambda \left. \left(\liminf_{r \to 0^+} \frac{1}{\mathrm{vol}(rB_n)} \int_{y + rB_n} f(t)\, \mathrm{d}t \right)^p \right]^{1/p} = \left((1 - \lambda)\phi(x)^p + \lambda\phi(y)^p \right)^{1/p}.$$

So we have that ϕ^p is concave and thus, since Ω is open, $\phi(x) \neq +\infty$ for all $x \in \Omega$. Hence, we may assert that $\phi : \Omega \longrightarrow \mathbb{R}_{\geq 0}$ is continuous (see e.g. [17, Theorem 10.1]). □

Proof of Theorem 1.2 If $\Omega = \mathbb{R}^n$ the statement follows from Lemma 4.1 and Theorem 3.2. For $p = 1/n$, in the same way, the result comes from Lemma 4.1 and Theorem 3.4 (also from Theorem 3.3 if $n = 1$). □

One can easily check that a nonzero measure μ satisfying (1.8), for any pair of balls contained in an open set A, fulfills $A \subset \mathrm{supp}(\mu)$. Indeed, given $x_0 \in \mathrm{supp}(\mu)$ (we notice that $\mathrm{supp}(\mu) \neq \emptyset$ because μ is nonzero), for any $x \in A$, $x \neq x_0$, there exist $y \in A$ and $\lambda \in (0, 1)$ such that $(1 - \lambda)x_0 + \lambda y = x$, since A is open. Thus, taking $K = x_0 + rB_n$ and $L = y + rB_n$ for $r > 0$ small enough, from (1.8) we have that $x \in \mathrm{supp}(\mu)$.

Now, as a consequence of Theorem 1.2, we get the following result for p-concave measures with arbitrary support.

Theorem 4.5 *Let $p > 0$ and let μ be a nonzero locally finite Borel measure on $\mathbb{R}^n$. Let $X = \operatorname{supp}(\mu)$, $H = \operatorname{aff} X$ and let $m = \dim X$. Suppose that μ is such that*

$$\mu((1 - \lambda)K + \lambda L) \geq M_p\big(\mu(K), \mu(L), \lambda\big) \tag{4.3}$$

holds for any pair of balls $K, L \in \mathcal{B}^n$, and all $\lambda \in (0, 1)$. Then $\mu_{|_X} = c\operatorname{vol}_m$ for some (constant) $c > 0$ if either $X = H$ or $p = 1/m$.

Proof Assume first that $m = n$ and let $\Omega = \operatorname{int} X$. By (4.3) and the definition of support, X is clearly convex. In particular, Ω is an open convex subset of $\mathbb{R}^n$ for which $\mu((1-\lambda)K + \lambda L)^p \geq (1-\lambda)\mu(K)^p + \lambda\mu(L)^p$ for any pair of balls $K, L \in \mathcal{B}^n$ with $K, L \subset \Omega$, and all $\lambda \in (0, 1)$. Hence, by Theorem 1.2, $\mu_{|_\Omega} = c\operatorname{vol}_n$ for some (constant) $c > 0$. Without loss of generality we may assume that $c = 1$ (otherwise we would work with the measure μ/c).

Thus, if $X = \mathbb{R}^n$, we are done. We next consider the case $X \neq \mathbb{R}^n$ and $p = 1/n$. Since the boundary of a convex set has Lebesgue measure zero (see e.g. [11]), and together with the already proved fact that $\mu_{|_\Omega} = \operatorname{vol}_n$, it is enough to show that $\mu(\operatorname{bd} X) = 0$ (we notice that X is a closed convex set with nonempty interior Ω and then $X = \Omega \cup \operatorname{bd} X$). To this end, by compactness arguments and by means of the relation

$$\mu(\operatorname{bd} X) = \mu\left(\bigcup_{k=1}^{+\infty} \big((\operatorname{bd} X) \cap kB_n\big)\right) = \lim_k \mu\big((\operatorname{bd} X) \cap kB_n\big),$$

it is enough to show that for each $x \in \operatorname{bd} X$ there exists $r = r_x > 0$ such that $\mu\big((x + rB_n) \cap \operatorname{bd} X\big) = 0$.

Suppose by contradiction that there exists $x \in \operatorname{bd} X$ such that

$$\mu\big((x + rB_n) \cap \operatorname{bd} X\big) > 0 \tag{4.4}$$

for all $r > 0$. Let $x_0 \in \Omega$ and let $r_0 > 0$ such that $x_0 + r_0B_n \subset \Omega$ and $(x + x_0)/2 + r_0B_n \subset \Omega$. Let $K = (x + r_0B_n) \cap X$ and $L = (x_0 - x) + K \subset x_0 + r_0B_n \subset \Omega$. Then $(K + L)/2 \subset (x + x_0)/2 + r_0B_n \subset \Omega$ and so, by (4.4) and the (equality case of the) Brunn-Minkowski inequality (1.1), we have

$$\mu\left(\frac{K+L}{2}\right) = \operatorname{vol}\left(\frac{K+L}{2}\right) = M_{1/n}\big(\operatorname{vol}(K), \operatorname{vol}(L), 1/2\big)$$

$$< M_{1/n}\big(\operatorname{vol}(K) + \mu(K \cap \operatorname{bd} X), \operatorname{vol}(L), 1/2\big) = M_{1/n}\big(\mu(K), \mu(L), 1/2\big),$$

a contradiction with (4.3). We point out that $(K + L)/2$ is measurable because it is convex, since it is a convex combination of convex sets.

Now the general case $m \leq n$ follows from the n-dimensional one because $\mu(\mathbb{R}^n \setminus X) = 0$ (cf. Definition 2.1) and thus (4.3) holds for any pair of balls $x + K$, $y + L$, with $x, y \in H$ and $K, L \in \mathcal{B}^m$, $K, L \subset H$, and all $\lambda \in (0, 1)$. □

Proof of Theorem 1.3 Let $x_0 \in \Omega$ be fixed. Since Ω is open, and $0 \in \Omega$, there exists $r_0 > 0$ such that $r_0 B_n \subset \Omega$ and $x_0 + r_0 B_n \subset \Omega$.

On the one hand, for all $u \in r_0 B_n$ and all $r > 0$ small enough, we get

$$\mu(x_0 + u + 2rB_n) \geq \left(\mu(x_0 + rB_n)^p + \mu(u + rB_n)^p\right)^{1/p}.$$

Thus, dividing by $\mathrm{vol}(rB_n)$, we have

$$\frac{2^n \mu(x + 2rB_n)}{\mathrm{vol}(2rB_n)} \geq \frac{\mu_s(x_0 + rB_n)}{\mathrm{vol}(rB_n)}$$

for all $x \in (x_0 + r_0 B_n) \subset \Omega$, where $\mathrm{d}\mu(x) = f(x)\mathrm{d}x + \mathrm{d}\mu_s(x)$ is the Lebesgue decomposition of μ.

So, by the above expression (and following the same steps to the proof of Lemma 4.2), one may conclude that μ is absolutely continuous.

Now, on the other hand, let $\phi : \Omega \longrightarrow \mathbb{R}_{\geq 0} \cup \{+\infty\}$ be the function given by

$$\phi(x) = \liminf_{r \to 0^+} \frac{1}{\mathrm{vol}(rB_n)} \int_{x+rB_n} f(t)\, \mathrm{d}t.$$

Then, following similar steps to the proof of Lemma 4.4 and taking $K = x_0 + (1 - \lambda)rB_n$, $L = u + \lambda rB_n$ for $\lambda \in (0, 1)$ and $r > 0$ small enough, we obtain

$$\phi(x_0 + u) \geq \left(\left((1 - \lambda)^n \phi(x_0)\right)^p + \left(\lambda^n \phi(u)\right)^p\right)^{1/p} \geq (1 - \lambda)^n \phi(x_0).$$

Taking limits as $\lambda \to 0^+$ we may assert that $\phi(x) \geq \phi(x_0)$ for all $x \in x_0 + r_0 B_n$. Exchanging the roles of x and x_0 we have that ϕ is constant on $x_0 + r_0 B_n$. Since x_0 is arbitrary, we get that ϕ is constant on every compact subset $C \subset \Omega$ and thus ϕ is so on Ω. The proof is now concluded because $\phi(x) = f(x)$ for almost every $x \in \Omega$ (by Theorem F).   □

Remark 4.6 Let E be a convex body with nonempty interior. The role played by $\mathcal{B}^n$ along this paper can be replaced by $\mathcal{F}^n = \{x + rE : x \in \mathbb{R}^n, r > 0\}$ since all the tools here involved are also true when exchanging the Euclidean unit ball B_n by E (see e.g. [18, Definition 7.9] and consequent results).

Acknowledgements The author would like to thank the anonymous referee for the careful reading of the paper and very useful suggestions which significantly improved the presentation.

This work was partially supported by "Programa de Ayudas a Grupos de Excelencia de la Región de Murcia", Fundación Séneca, 19901/GERM/15 and by ICMAT Severo Ochoa project SEV-2011-0087 (MINECO).

References

1. F. Barthe, Autour de l'inégalité de Brunn-Minkowski. Ann. Fac. Sci. Toulouse Math. Ser. 6 **12**(2), 127–178 (2003)
2. C. Borell, Convex set functions in d-space. Period. Math. Hung. **6**, 111–136 (1975)
3. H.J. Brascamp, E.H. Lieb, On extensions of the Brunn-Minkowski and Prékopa-Leindler theorems, including inequalities for log concave functions and with an application to the diffusion equation. J. Funct. Anal. **22**(4), 366–389 (1976)
4. P.S. Bullen, *Handbook of Means and Their Inequalities*. Mathematics and Its Applications, vol. 560 (Kluwer Academic Publishers Group, Dordrecht, 2003). Revised from the 1988 original
5. D.L. Cohn, *Measure Theory*, 2nd edn. (Birkhäuser Boston Inc., Boston, 2013)
6. S. Dharmadhikari, K. Joag-Dev, *Unimodality, Convexity, and Applications*. Probability and Mathematical Statistics (Academic, Boston, MA, 1988)
7. R.J. Gardner, The Brunn-Minkowski inequality. Bull. Am. Math. Soc. **39**(3), 355–405 (2002)
8. R.J. Gardner, A. Zvavitch, Gaussian Brunn-Minkowski inequalities. Trans. Am. Math. Soc. **362**(10), 5333–5353 (2010)
9. G.H. Hardy, J.E. Littlewood, G. Pólya, *Inequalities*. Cambridge Mathematical Library (Cambridge University Press, Cambridge, 1988). Reprint of the 1952 edition
10. M.A. Hernández Cifre, J. Yepes Nicolás, Refinements of the Brunn-Minkowski inequality. J. Convex Anal. **21**(3), 727–743 (2014)
11. R. Lang, A note on the measurability of convex sets. Arch. Math. **47**(1), 90–92 (1986)
12. L. Leindler, On certain converse of Hölder's inequality II. Acta Math. Sci. (Szeged) **33**, 217–223 (1972)
13. G. Livshyts, A. Marsiglietti, P. Nayar, A. Zvavitch, On the Brunn-Minkowski inequality for general measures with applications to new isoperimetric-type inequalities. Trans. Am. Math. Soc. (2015). arXiv:1504.04878 [math.PR]
14. A. Marsiglietti, On the improvement of concavity of convex measures. Proc. Am. Math. Soc. **144**(2), 775–786 (2016)
15. A. Prékopa, Logarithmic concave measures with application to stochastic programming. Acta Math. Sci. (Szeged) **32**, 301–315 (1971)
16. A.W. Roberts, D.E. Varberg, *Convex Functions*. Pure and Applied Mathematics, vol. 57 (Academic Press [A subsidiary of Harcourt Brace Jovanovich, Publishers], New York, London, 1973)
17. R.T. Rockafellar, *Convex Analysis* (Princeton University Press, Princeton, NJ, 1970)
18. W. Rudin, *Real and Complex Analysis*, 3rd edn. (McGraw-Hill, New York, 1987)
19. R. Schneider, *Convex Bodies: The Brunn-Minkowski Theory*, 2nd edn. (Cambridge University Press, Cambridge, 2014)

Printed in the United States
By Bookmasters